General Safe Practices for Working with Engineered Nanomaterials in Research Laboratories

DEPARTMENT OF HEALTH AND HUMAN SERVICES
Centers for Disease Control and Prevention
National Institute for Occupational Safety and Health

This document is in the public domain and may be freely copied or reprinted.

Disclaimer

Mention of any company or product does not constitute endorsement by the National Institute for Occupational Safety and Health (NIOSH). In addition, citations to Web sites external to NIOSH do not constitute NIOSH endorsement of the sponsoring organizations or their programs or products. Furthermore, NIOSH is not responsible for the content of these Web sites. All Web addresses referenced in this document were accessible as of the publication date.

Ordering Information

To receive documents or other information about occupational safety and health topics, contact NIOSH at

> Telephone: **1-800-CDC-INFO** (1-800-232-4636)
> TTY: 1-888-232-6348
> E-mail: cdcinfo@cdc.gov
>
> or visit the NIOSH Web site at **www.cdc.gov/niosh**.

For a monthly update on news at NIOSH, subscribe to *NIOSH eNews* by visiting **www.cdc.gov/niosh/eNews**.

DHHS (NIOSH) Publication No. 2012-147

May 2012

SAFER • HEALTHIER • PEOPLE™

Foreword

The National Institute for Occupational Safety and Health (NIOSH) is pleased to present *General Safe Practices for Working with Engineered Nanomaterials in Research Laboratories*. Engineered nanomaterial applications are rapidly expanding throughout the United States and worldwide. The research community is at the front line of creating these new nanomaterials, testing their usefulness in a variety of applications, and determining their toxicological and environmental impacts.

With the publication of this document, NIOSH hopes to raise awareness of the occupational safety and health practices that should be followed during the synthesis, characterization, and experimentation with engineered nanomaterials in a laboratory setting. The document contains recommendations on engineering controls and safe practices for handling engineered nanomaterials in laboratories and some pilot scale operations. This guidance was designed to be used in tandem with well-established practices and the laboratory's chemical hygiene plan. As our knowledge of nanotechnology increases, so too will our efforts to provide additional guidance materials for working safely with engineered nanomaterials.

/s
John Howard, M.D.
Director, National Institute for
   Occupational Safety and Health
Centers for Disease Control and Prevention

PAGE LEFT INTENTIONALLY BLANK

Acknowledgments

This document is based on input from several subject matter experts and was initiated as a joint effort under a Memorandum of Understanding between NIOSH and the Center for High-rate Nanomanufacturing (CHN). Some of the specific content was derived from a report generated by Michael Ellenbecker and Su-Jung (Candace) Tsai at the University of Massachusetts Lowell (UMass Lowell), one of the CHN member campuses, and was supported by a contract from the NIOSH Nanotechnology Research Center (NTRC). Paul Schulte is the manager and Charles Geraci is the coordinator of the NIOSH nanotechnology cross-sector program. Special thanks go to Catherine Beaucham and Laura Hodson for writing and organizing this report. Others who contributed substantially to the writing and research include Mark Hoover and Ralph Zumwalde.

The NIOSH NTRC also acknowledges the contributions of Gino Fazio for desktop publishing and graphic design, Michael Elliot and Terri Pearce for review, and John Lechliter and Seleen Collins for editing the report. Photographs are courtesy of Catherine Beaucham and Mark Methner of NIOSH, Michael Ellenbecker and Su-Jung (Candace) Tsai of UMass Lowell and Mia Ertas of the University of Albany College of Nanoscale Science & Engineering (CNSE).

External Expert Peer Review

Lawrence M. Gibbs, MPH, CIH
Associate Vice Provost for EH&S
Stanford University

Bruce C. Stockmeier, CIH
ES&H Coordinator
Argonne National Laboratory
Center for Nanoscale Materials

William Kojola
Industrial Hygienist
The American Federation of Labor—
  Congress of Industrial Organizations
  (AFL-CIO)

FuAnjali Lamba, MPH, CIH
Senior Industrial Hygienist
Chemical Engineering Branch
Office of Pollution Prevention and Toxics
U.S. Environmental Protection Agency

PAGE LEFT INTENTIONALLY BLANK

Executive Summary

Nanotechnology, the manipulation of matter at a nanometer scale to produce new materials, structures, and devices having new properties, may revolutionize life in the future. It has the potential to impact medicine through improved disease diagnosis and treatment technologies and to impact manufacturing by creating smaller, lighter, stronger, and more efficient products. Nanotechnology could potentially decrease the impact of pollution by improving methods for water purification or energy conservation. Although engineered nanomaterials present seemingly limitless possibilities, they bring with them new challenges for identifying and controlling potential safety and health risks to workers. Of particular concern is the growing body of evidence that occupational exposure to some engineered nanomaterials can cause adverse health effects.

As with any new technology or new material, the earliest exposures will likely occur for those workers conducting discovery research in laboratories or developing production processes in pilot plants. The research community is at the front line of creating new nanomaterials, testing their usefulness in a variety of applications and determining their toxicological and environmental impacts. Researchers handling engineered nanomaterials in laboratories should perform that work in a manner that protects their safety and health. This guidance document provides the best information currently available on engineering controls and safe work practices to be followed when working with engineered nanomaterials in research laboratories.

Risk Management

Risk management is an integral part of occupational health and safety. Potential exposures to nanomaterials can be controlled in research laboratories through a flexible and adaptive risk management program. An effective program provides the framework to anticipate the emergence of this technology into laboratory settings, recognize the potential hazards, evaluate the exposure to the nanomaterial, develop controls to prevent or minimize exposure, and confirm the effectiveness of those controls.

Hazard Identification

Experimental animal studies indicate that potentially adverse health effects may result from exposure to nanomaterials. Experimental studies in rodents and cell cultures have shown that the toxicity of ultrafine particles or nanoparticles is greater than the toxicity of the same mass of larger particles of similar chemical composition.

Research demonstrates that inhalation is a significant route of exposure for nanomaterials. Evidence from animal studies indicates that inhaled nanoparticles may deposit deep in lung tissue, possibly interfering with lung function. It is also theorized that nanoparticles may enter the bloodstream through the lungs and transfer to other organs. Dermal exposure and subsequent penetration of nanomaterials may cause local or systemic effects. Ingestion is a third potential route of exposure. Little is known about the possible adverse effects of ingestion of nanomaterials, although some evidence suggests that nanosized particles can be transferred across the intestinal wall.

Exposure Assessment

Exposure assessment is a key element of an effective risk management program. The exposure assessment should identify tasks that contribute to nanomaterial exposure and the workers conducting those tasks. An inventory of tasks should be developed that includes information on the duration and frequency of tasks that may result in exposure, along with the quantity of the material being handled, dustiness of the nanomaterial, and its physical form. A thorough understanding of the exposure potential will guide exposure assessment measurements, which will help determine the type of controls required for exposure mitigation.

Exposure Control

Exposure control is the use of a set of tools or strategies for decreasing or eliminating worker exposure to a particular agent. Exposure control consists of a standardized hierarchy to include (in priority order): elimination, substitution, isolation, engineering controls, administrative controls, or if no other option is available, personal protective equipment (PPE).

Substitution or elimination is not often feasible for workers performing research with nanomaterials; however, it may be possible to change some aspects of the physical form of the nanomaterial or the process in a way that reduces nanomaterial release.

Isolation includes the physical separation and containment of a process or piece of equipment, either by placing it in an area separate from the worker or by putting it within an enclosure that contains any nanomaterials that might be released.

Engineering controls include any physical change to the process that reduces emissions or exposure to the material being contained or controlled. Ventilation is a form of engineering control that can be used to reduce occupational exposures to airborne particulates. General exhaust ventilation (GEV), also known as dilution ventilation, permits the release of the contaminant into the workplace air and then dilutes the concentration to an acceptable level. GEV alone is not an appropriate control for engineered nanomaterials or any other uncharacterized new chemical entity. Local exhaust ventilation (LEV), such as the standard laboratory chemical hood (formerly known as a laboratory fume hood), captures emissions at the source and thereby removes contaminants from the immediate occupational environment. Using selected forms of LEV properly is appropriate for control of engineered nanomaterials.

Administrative controls can limit workers' exposures through techniques such as using job-rotation schedules that reduce the time an individual is exposed to a substance. Administrative controls may consist of standard operating procedures, general or specialized housekeeping procedures, spill prevention and control, and proper labeling and storage of nanomaterials. Employee training on the appropriate use and handling of nanomaterials is also an important administrative function.

PPE creates a barrier between the worker and nanomaterials in order to reduce exposures. PPE may include laboratory coats, impervious clothing, closed-toe shoes, long pants, safety glasses, face shields, impervious gloves, and respirators.

Other Considerations

Control verification or confirmation is essential to ensure that the implemented tools or strategies are performing as specified. Control verification can be performed with traditional industrial hygiene sampling methods, including area sampling, personal sampling, and real-time measurements. Control verification may also be achieved by monitoring the performance parameters of the control device to ensure that design and performance criteria are met.

Other important considerations for effective risk management of nanomaterial exposure include fire and explosion control. Some studies indicate that nanomaterials may be more prone to explosion and combustion than an equivalent mass concentration of larger particles.

Occupational health surveillance is used to identify possible injuries and illnesses and is recommended as a key element in an effective risk management program. Basic medical screening is prudent and should be conducted under the oversight of a qualified healthcare professional.

PAGE LEFT INTENTIONALLY BLANK

Contents

Foreword iii
Acknowledgments v
Executive Summary vii
Abbreviations xiii
1 Introduction 1
2 Scope 1
3 Risk Management 2
4 Hazard Identification 3
5 Exposure Assessment 5
    5.1 Safety Through the Life Cycle of a Nanomaterial 6
        5.1.1 Synthesis 8
        5.1.2 Characterization and Purification 9
        5.1.3 Application and Material Testing 9
6 Recommendations for Exposure Control 10
    6.1 Elimination or Substitution 10
    6.2 Isolation and Engineering Controls 12
        6.2.1 Containment 13
        6.2.2 Ventilation 13
    6.3 Administrative Controls 14
        6.3.1 Employee Training 14
        6.3.2 Labeling and Storage 15
    6.4 Personal Protective Equipment 15
        6.4.1 Protective Clothing 16
        6.4.2 Respirators 17
7 Local Exhaust Ventilation 18
    7.1 Laboratory Chemical Hoods 19
        7.1.1 Working with Nanomaterial Powders in Chemical Hoods 24
        7.1.2 New Hood Designs 24
    7.2 Alternatives to Conventional Chemical Hoods 24
        7.2.1 Glovebox Enclosures 24

7.2.2 Biological Safety Cabinets	25
7.2.3 Powder Handling Enclosures	27
8 Methods for Exposure Control Verification	28
9 Periodic Re-evaluations of the Risk Management Program	29
10 Guidance on Developing a Control Scheme (Control Banding)	31
11 Fire and Explosion Control	33
12 Management of Nanomaterial Spills	34
13 Occupational Health Surveillance	34
14 Conclusions	35
References	37

Abbreviations

ACGIH	American Conference of Governmental Industrial Hygienist
AIHA	American Industrial Hygiene Association
ANSI	American National Standards Institute
ASHRAE	American Society of Heating, Refrigeration, and Air Conditioning Engineers
BSC	biological safety cabinet
CFR	Code of Federal Regulations
CHN	Center for High-rate Nanomanufacturing
CNT	carbon nanotube
DNA	deoxyribonucleic acid
FLAR	flame aerosol reactor
FSP	flame spray pyrolysis
GEV	general exhaust ventilation
HEPA	high-efficiency particulate air
HVAC	heating, ventilation, and air conditioning
ICRP	International Commission on Radiological Protection
ISO	International Organization for Standardization
LEV	local exhaust ventilation
LLNL	Lawrence Livermore National Laboratory
μm	micrometer
MWCNT	multi-walled carbon nanotube
nm	nanometer
NAS	National Academy of Sciences
NIOSH	National Institute for Occupational Safety and Health
NTRC	Nanotechnology Research Center
OPC	optical particle counter
OECD	Organization for Economic Co-operation and Development
OSHA	Occupational Safety and Health Administration
PEL	permissible exposure limit
PtD	Prevention through Design
REL	recommended exposure limit
RL	risk level
SWCNT	single-walled carbon nanotube

PAGE LEFT INTENTIONALLY BLANK

1 Introduction

According to The International Organization for Standardization Technical Committee 229 (Nanotechnologies) (ISO/TS 27687:2008), a nano-object is a material with one, two, or three external dimensions in the 1- to 100-nm size range. Nano-objects are frequently incorporated into a larger matrix known as a nanomaterial. Nanoparticles are a specific type of nano-object, with all three external dimensions at the nanoscale. An additional term, ultrafine particles, is used to describe nanometer-diameter particles that have not been intentionally produced but are the incidental products of processes [NIOSH 2009a]. For purposes of this document, the term *nanomaterial* is used to describe engineered nano-objects, including engineered nanoparticles.

Nanomaterials are increasingly being used in optoelectronic, electronic, magnetic, medical imaging, drug delivery, cosmetic, catalytic, and other applications. Although nanomaterials present seemingly limitless possibilities, they bring with them new challenges to understanding, predicting, and managing potential safety and health risks to workers. Exposures to nanomaterials can involve a wide range of nanomaterial sizes, shapes, functionalities, concentrations, chemical compositions, and exposure frequencies or durations. Researchers working with engineered nanomaterials have the potential to be exposed through a variety of sources and processes, including leaks from equipment used in the synthesis of nanomaterials, manipulating dry nanopowders, sonicating liquid suspensions, or mechanically disrupting materials containing or coated with nanomaterials [Aitken et al. 2004; Johnson et al. 2010]. A growing body of evidence indicates that exposure to some of these engineered nanomaterials can cause adverse health effects. Based on this preliminary toxicological data, prudent practice dictates controlling occupational exposure to nanomaterials.

2 Scope

As with any new technology, the earliest exposures will likely occur among those workers conducting research in laboratories and pilot plants. Researchers handling engineered nanomaterials in laboratories and pilot scale operations should perform that work in a manner that is protective of their safety and health. Although incidental nanoparticles (also known as ultrafine particles) exist in nature, the focus of this document is to provide guidance on the safe handling of purposely designed, engineered nanomaterials in research laboratories. The information may also be applicable in some pilot-scale facilities.

Research laboratories include any facility performing basic or applied research involving nanomaterials. Nanomaterial research laboratories may be housed at universities, government agencies, and private companies. Research laboratories may produce their own nanomaterials, work with nanomaterials produced by others, or some combination of both. Laboratory-scale production typically consists of relatively small amounts of nanomaterial, ranging from a few milligrams for highly sophisticated materials such as quantum dots to a few kilograms for less-sophisticated materials such as metal oxides. Laboratories conducting applied research may also produce materials on a pilot scale, which typically increases material volumes by a factor of 10 or more. Pilot-scale equipment is generally similar to industrial-scale processes, but it produces much smaller quantities of nanomaterial.

3 Risk Management

Exposures to engineered nanomaterials can be controlled in the research laboratory by a comprehensive risk management program that includes task hazard/risk analysis, engineering controls, administrative controls, and use of PPE. Implementing an effective program should address the following elements of hazard surveillance.

Hazard Identification: Is there reason to believe that the nanomaterial of interest could be harmful?

Exposure Assessment: Is there potential for exposure to the nanomaterial or other chemical or physical hazards?

Exposure Control: What procedures are in place or should be developed to minimize or eliminate worker exposure(s)?

The answers to these questions will help to formulate a program that includes the following:

- A written health and safety policy covering all types of chemical and physical hazards in the workplace, in accordance with the U.S. regulatory requirement 29 CFR 1910.1450, the Occupational Safety and Health Administration's (OSHA's) laboratory standard, including development of a Chemical Hygiene Plan.
- A clear delineation of roles and responsibilities for everyone involved in laboratory or pilot plant research.
- Effective procedures for documentation, communication, and employee training.
- Incorporation of input from safety professionals, industrial hygienists, and occupational health professionals, as appropriate.

Figure 1 illustrates components of an overall health and safety program that includes nanomaterial risk management [Schulte et al. 2008a]. Additional guidance on prudent practices in the laboratory [NRC 2011] can be obtained from the National Academy of Sciences (NAS).

Figure 1. Components of an overall health and safety program. Modified from Schulte et al. [2008].

4 Hazard Identification

The unique properties of materials at the nanoscale have raised concerns regarding health effects that might result from occupational exposures. The toxicity of a nanomaterial will be a function of its substance-specific toxicity, as influenced by physicochemical characteristics (including those unique to the nanoscale form of the substance) and contaminants [Trout and Schulte 2010].

Results of studies in which animals and humans were exposed to ultrafine or other respirable particles provide a basis of concern for possible adverse health effects due to engineered nanomaterial exposures. Experimental studies in rodents and cell cultures have shown that the toxicity of ultrafine or nanoparticles is greater than that of the same mass of larger particles of similar chemical composition [Oberdörster et al. 1992; Oberdörster et al. 1994; Lison et al. 1997; Tran et al. 1999; Tran et al. 2000; Brown et al. 2001; Barlow et al. 2005; Duffin et al. 2007]. In addition to particle size and surface area, other particle characteristics may influence toxicity, including surface functional groups or coatings, solubility, shape, and the ability to generate reactive oxygen species [Duffin et al. 2002; Maynard and Kuempel 2005; Oberdörster et al. 2005; Donaldson et al. 2006].

Several articles have investigated the toxicity of carbon nanotubes (CNTs) in experimental animal studies [Lam et al. 2004; Shvedova et al. 2005; Donaldson et al. 2006;

Lam et al. 2006; Kisin et al. 2007; Li et al. 2007; Kane and Hurt 2008; Miyawaki et al. 2008; Poland et al. 2008; Shvedova et al. 2008; Erdely et al. 2009; Ma-Hock et al. 2009; Shvedova et al. 2009; Pauluhn 2010]. The results from these studies indicate potential respiratory health risks from exposure to CNTs, including granulomatous pneumonia and fibrosis. Evidence also indicates that when multi-walled carbon nanotubes (MWCNTs) are administered intraperitoneally to mice, the MWCNTs have asbestos-like pathogenicity [Poland et al. 2008; Takagi et al. 2008]. Although a causal link has not been established, there is concern about possible cancer hazards in addition to potential for fibrosis/nonmalignant respiratory disease.

Additional studies have investigated the DNA damage caused by nanosized metals and metal oxides [Karlsson et al. 2009; Singh et al. 2009]. Although it cannot be concluded that metal oxide nanoparticles are always more toxic than their micrometer counterparts, nanosized copper oxide (CuO) was found to be much more toxic than micrometer-sized CuO [Karlsson et al. 2009].

Inhalation is considered the primary route of potential exposure in the nanomaterial workplace. Evidence indicates discrete nanoparticles are deposited in the lungs to a greater extent than larger respirable particles [ICRP 1994]. Some nanoparticles are thought to enter the bloodstream from the lungs and then transfer to other organs [Takenaka et al. 2001; Nemmar et al. 2002; Oberdörster et al. 2002; Geiser et al. 2005]. It is further postulated that some nanomaterials may move from the nose to the brain though the blood-brain barrier [Oberdörster et al. 2004; Elder et al. 2006].

Dermal exposure to nanomaterials is also a potential exposure pathway. Possible harmful effects may occur locally, or the substances may be absorbed through the skin and cause systemic effects. Studies indicate particles smaller than 1 μm in diameter may penetrate intact skin [Tinkle et al. 2003; Ryman-Rasmussen et al. 2006]. Dermal irritation has been seen following topical application of single-walled carbon nanotubes (SWCNTs) to nude mice [Shvedova et al. 2003; Murray et al. 2007], although it is not known whether skin penetration could occur and result in adverse health effects. Additional data are needed to extrapolate these findings for identifying any occupational health risks and for investigating the dermal toxicity of other nanomaterials.

Ingestion of nanomaterials might occur due to unintentional hand-to-mouth contact, thereby allowing possible transfer to other body organs via the gastrointestinal tract. The mucociliary escalator system, where particles that are deposited in the lung are transferred by coughing to the pharynx and subsequently swallowed, is an additional path to ingestion. Little is known about possible adverse effects from ingestion of nanomaterials; however, some evidence indicates smaller particles can be transferred across the intestinal wall more readily than larger particles [Behrens et al. 2002].

5 Exposure Assessment

An exposure assessment should identify tasks that may expose workers to nanomaterials and also identify the researchers conducting those tasks. Such an assessment would review the process and material flow plans for the facility and the status of specific projects. It would include staff interviews and a walk-through of the facility (laboratory) to ensure that all activities and potential exposure pathways are identified. The inventory of tasks and workers should include information on the potential magnitude, duration, and frequency of exposure during different job tasks, or at specific processes, and the amounts of materials being used. Current work practices and existing engineering controls should be evaluated.

The work tasks should be inventoried and prioritized according to the potential for occupational exposure. Examples of tasks and product activities include the following:

- Material receipt, unpacking, and delivery.
- Laboratory operations (synthesis, analytical, and quality assurance activities).
- Cleaning and maintenance.
- Storage, packaging, and shipping.
- Reasonably foreseeable emergencies.
- Waste management.

Determinants of potential exposure to nanomaterials may include dustiness, type of process, quantity of material handled, and duration and frequency of employee exposure. These elements are summarized below and should be taken into account when implementing exposure control measures.

Dustiness

The dustiness of the nanomaterial can influence potential exposures and the selection of the appropriate engineering control. Dustiness describes the tendency of particles to become and/or stay airborne and refers not only to the physical form of the nanomaterial but also to the electrostatic repulsive forces inherent in the particle. For example, the "dustiness" of the nanomaterial is influenced by its particle bulk density and morphology (shape, diameter, and length), as well as the incorporation of the nanomaterial into slurries or liquid suspensions. Nanomaterials in dry powder form tend to pose the greatest risk for inhalation exposure, while nanomaterials suspended in a liquid typically present less risk via inhalation. Exceptions have been identified during some laboratory processes such as sonication, which resulted in an increase in airborne nanomaterials [Johnson et al. 2010]. Electrostatic forces influence the stability of particle dispersion in air. These electrostatic forces therefore affect dustiness and should be controlled where possible. Nanomaterials with little or no repulsive forces will tend to be more likely to form aggregates and therefore be less dusty. Nanomaterials incorporated into a solid matrix present the least risk for inhalation exposure because of their limited mobility as long as they are maintained within the matrix.

Process

Some material handling, synthesis, and manufacturing processes can increase the risk of employee exposure. Open, manual handling of bulk nanomaterials, as well as high-energy processes such as milling, sonication, grinding, and high-speed blending, could cause the release of nanomaterials [Gohler et al. 2010; Johnson et al. 2010]. Consideration should also be given to the possibility of intentional or inadvertent chemical changes during a work task that may alter the toxicity of a nanomaterial.

Quantity, Duration, and Frequency of Task

The quantity of the nanomaterial that is synthesized, received, or handled in the laboratory will significantly influence exposure potential. Research laboratories may handle quantities ranging from milligrams to several grams or even kilograms of a nanomaterial. As quantities increase, consideration of additional control measures may be required. Exposure potential may be influenced by the duration and frequency of the task(s). Small quantities used on an infrequent basis may not require the same level of control measure that large quantities used daily would require.

Engineering controls should be the primary means of controlling exposures, except in situations (e.g., emergencies) where such controls may not be feasible. In those circumstances, other control measures may be required (e.g., respirator use).

5.1 Safety Through the Life Cycle of a Nanomaterial

To ensure the health and safety of those working with nanomaterials, the exposure sources during the nanomaterial product life cycle should be evaluated. Exposure sources include nanomaterial synthesis reactors, nanoparticle collection and handling, product fabrication with nanomaterials, product use, and product disposal [Sahu and Biswas 2010]. Table 1 contains some selected activities with potential exposure sources and recommended engineering controls. The ultimate disposal of the nanomaterial and contaminated refuse should follow all applicable federal, state, and local regulations.

Consideration should be given to installing high-efficiency particulate air (HEPA) filters on laboratory chemical hoods or other individual exhaust duct systems. The decision to use HEPA filtration should be based on evaluation of the contaminant characteristics, maintenance and protection of the fan motor and other exhaust parts, energy

Table 1. Employee activities and recommended minimum controls.

State of the nanomaterial	Employee activity	Potential exposure source	Recommended engineering controls
Bound or fixed nanostructures (polymer matrix)	• Mechanical grinding, alloying, etching, lithography, erosion, mechanical abrasion, grinding, sanding, drilling, heating, cooling	• Nanomaterials may be released during grinding, drilling, and sanding. Heating or cooling may damage the matrix, allowing release of nanomaterial.	• Local exhaust ventilation • Laboratory chemical hood (with HEPA-filtered exhaust) • HEPA-filtered exhausted enclosure (glovebox) • Biological safety cabinet class II type A1, A2, vented via thimble connection, or B1 or B2
Liquid suspension, liquid dispersion	• Synthesis methods: chemical precipitation, chemical deposition, colloidal, electrodeposition crystallization, laser ablation (in liquid) • Pouring and mixing of liquid containing nanomaterials • Sonication • Spraying • Spray drying	• Exposures may result from aerosolization of nanoparticles during sonication or spraying, equipment cleaning and maintenance, spills, or product recovery (dry powders).	• Laboratory chemical hood (with HEPA-filtered exhaust) • HEPA-filtered exhausted enclosure (glovebox) • Biological safety cabinet class II type A1, A2, vented via thimble connection, or B1 or B2
Dry dispersible nanomaterials and agglomerates	• Collection of material (after synthesis), material transfers, weighing of dry powders, mixing of dry powders	• Exposures may occur during any dry powder handling activity or product recovery.	• Laboratory chemical hood with HEPA-filtered exhaust • HEPA-filtered exhausted enclosure (glovebox) • Biological safety cabinet class II, B1 or B2
Nanoaerosols and gas phase synthesis (on substrate)	• Vapor deposition, vapor condensation, rapid solidification, aerosol techniques, gas phase agglomeration, inert gas condensation (flame pyrolysis, high temperature evaporation), or spraying	• Exposures may occur with direct leakage from the reactor, product recovery, processing and packaging of dry powder, equipment cleaning, and maintenance.	• Glovebox or other sealed enclosure with HEPA-filtered exhaust • Appropriate equipment for monitoring toxic gas (e.g., CO)

Table adapted from the summary of recommended nanomaterial controls from the University of New Hampshire [UNH 2009], the University of North Carolina [UNC 2011], and the Research Report 274 [Aiken et al. 2004] prepared by the Institute of Occupational Medicine for the Health and Safety Executive.

requirements, and any applicable environmental release rules. NIOSH recommends the use of HEPA filtration on local exhaust ventilation, laboratory chemical hoods, low-flow enclosures, and any other containment enclosures as a best practice during the handling of engineered nanomaterials.

5.1.1 Synthesis

Nanomaterial synthesis represents the first step in the nanomaterials exposure pathway. Numerous methods can be used to synthesize nanomaterials, and the nature of the potential exposure depends on the specific synthesis process and the stage within the process. Appropriate engineering controls will depend greatly on the synthesis method utilized and the step in the process where the exposure might occur.

Because of the possibility of equipment leaks, synthesis processes should be carried out in an isolated area or in an enclosure operating under negative pressure and exhausted through HEPA filters [Seaton et al. 2010]. Precautions such as local exhaust ventilation and PPE should be utilized when cleaning or performing maintenance on the equipment. Furthermore, general ventilation in the laboratory is often not sufficient to effectively clear nanomaterials released into the general room air over a 30-minute period; therefore, researchers should leave the hood fan on even after synthesis is complete [Sahu and Biswas 2010]. Finally, exposures could occur during product recovery,

Carry out operations in a manner that minimizes the risk of exposure to nanomaterials from inhalation or dermal contact. Principles that contribute to minimizing the risk of exposure to nanomaterials in the laboratory include the following:

- Handle nanomaterials in dry powder form with care to minimize the generation of airborne dust and to minimize dermal contact.
- Nanomaterials suspended in a liquid present less risk for becoming airborne than nanomaterials in dry powder form under normal handling conditions, but they may present a dermal risk, especially if the nanomaterial is suspended in a solvent.
- Nanomaterials suspended in a liquid may be aerosolized during certain handling activities (for example, during sonication).
- Nanomaterials incorporated into a solid matrix are least likely to become airborne because of their limited mobility. However, under certain circumstances these nanomaterials may still pose some risk, such as if the solid matrix is cut, sawed, drilled, sanded, or handled in any way that creates a dust or releases the nanomaterial.
- The quantity of material handled contributes greatly to the risk of exposure. Operations involving the use of nanomaterials should always use the minimum quantity required for the particular experiment or process.

packaging, and shipping phases and therefore should be identified and controlled according to the state of the nanomaterial during that stage of synthesis (Table 1).

5.1.2 Characterization and Purification

Once the nanomaterial has been synthesized, it may undergo characterization, purification, or other modification steps such as the addition of surface coatings to functionalize it. Safety precautions and standard operating procedures should be developed and followed for hazards associated with the characterization, purification, or functionalization of the nanomaterial.

Characterization includes the determination of the size and shape of the nanomaterial, atomic and electronic structures, and any other important chemical or physical properties [Rao and Biswas 2009]. This process may include various analytical methods such as microscopy, X-ray diffraction, and spectroscopy. Purification or processing of nanomaterials is used to remove impurities from the nanomaterial of interest. For example, a raw carbon nanotube material may contain the catalyst used in the synthesis process. Purification techniques include high-temperature heat treatments, the application of highly acidic or caustic substances, or the use of potentially hazardous solvents. Functionalization modifies the particle surface by attaching another substance, which may change the toxicity or behavior of the nanomaterial.

5.1.3 Application and Material Testing

The third phase of nanomaterial processing in research laboratories involves the application and testing of the nanomaterial or nanoenabled material. This may involve combining the nanomaterial into other matrices, applying nanomaterials onto surfaces, or destructive testing of substances containing nanomaterials.

Avoid manipulating nanomaterials in open systems or in a free particle state (e.g., handling dry nanopowders on a bench top).

- Preferably, (1) keep nanomaterials bound in a matrix, (2) keep them suspended in a liquid, (3) keep them sealed in a container, or (4) use appropriate engineering controls.
- For larger processes that cannot fit in a fume hood or glovebox (e.g., injection molding), control emissions with properly designed local exhaust ventilation.
- Transfer nanomaterial samples between workstations (such as exhaust hoods, gloveboxes, furnaces) in sealed, unbreakable, labeled containers.
- Avoid generating nanoparticle aerosols (e.g., through sonication) on bench tops. Use appropriate laboratory exhaust and containment systems.

The most stable form for most nanomaterials occurs when they are bound within a solid matrix. However, destructive treatment of the matrix, such as during grinding, sanding, and drilling, may lead to the release of nanoparticles or larger particles containing nanomaterial. Thermal stresses such as melting plastics may also cause nanomaterial release. Local exhaust ventilation (Section 7.0) should be used during destructive handling of matrices containing nanomaterials [UNH 2009].

Working with nanomaterials in liquid during activities where energy is applied such as sonication or mixing may generate the airborne release of respirable droplets containing nanomaterials [Johnson et al. 2010]. Proper controls should be used during these operations (Table 1).

6 Recommendations for Exposure Control

Among the most effective means to prevent occupational injuries and illnesses are anticipating potential occupational safety and health hazards early in the development of the technology or process and incorporating safe practices into all design, implementation, and operation phases. Prevention through Design (PtD) is a management tool for protecting workers from potentially unsafe work conditions. It emphasizes the importance of employee health and safety through the design, construction, manufacture, use, maintenance, and ultimate disposal or reuse of tools, equipment, machinery, substances, work processes, and work premises [NIOSH 2010b]. PtD addresses occupational safety and health needs by eliminating hazards and minimizing risks to workers throughout the life cycle of the process (Figure 2) [Schulte et al. 2008b]. Many nanotechnology research laboratories recognize PtD as a cost-effective means to enhance occupational safety and health and have incorporated PtD management practices within their facilities [Murashov and Howard 2009].

Prevention through Design strategies follow the standard hierarchy of controlling workplace hazards, which includes (1) eliminating, substituting, or modifying the nanomaterials; (2) engineering the process to minimize or eliminate exposure to the nanomaterials; (3) implementing administrative controls that limit the quantity or duration of exposure to the nanomaterials; and (4) providing for use of PPE.

6.1 Elimination or Substitution

For nanomaterial researchers, it is often not feasible to eliminate or substitute the nanomaterial. It may be possible, however, to change some aspects of the process in a way that reduces release of the nanomaterial. For example, working with nanomaterials suspended in a liquid is a significant improvement over working with them in dry powder form, because the potential for airborne release is reduced in most laboratory processes.

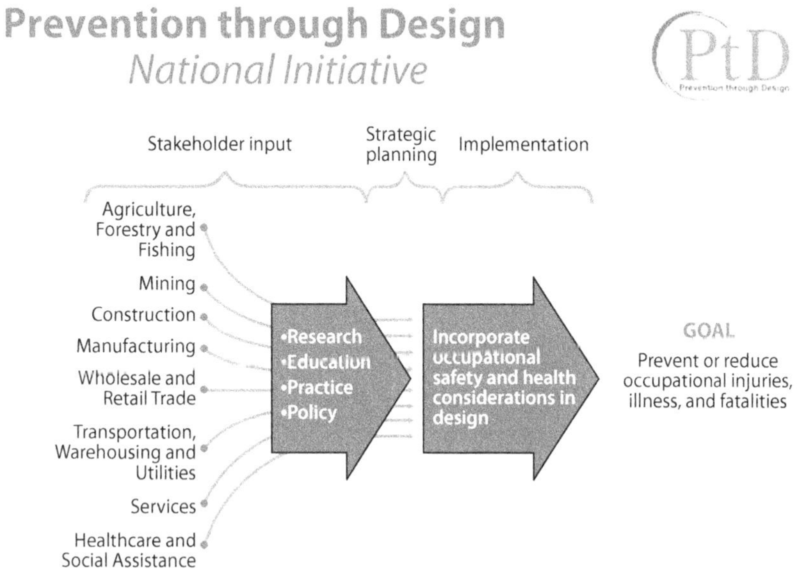

Prevention through Design
National Initiative

Figure 2. Prevention through Design

However, physical agitation of the liquid (e.g., sonication) may aerosolize small droplets containing the nanomaterial [Johnson et al. 2010] (Figure 3).

Opportunities for eliminating the use of hazardous materials or substituting for less hazardous forms do exist in other aspects of nanomaterials production. Engineered nanoparticle research often requires the use of solvents and other potentially hazardous chemicals. A recent article on optimizing the properties of carbon nanotubes (CNTs) reported that eight different solvents were evaluated for the optimization; all eight solvents, including the solvents benzene, toluene, ethyl acetate, and dimethylformamide, were considered toxic to different degrees [Ju et al. 2009]. Researchers should always attempt to identify and use chemical processes that utilize nontoxic or less-toxic alternatives whenever possible, in order to minimize worker exposures and environmental releases when the process is scaled up to full production. This control strategy, substituting a less toxic material in production processes, has been the focus of much research during the past 20 years. One source

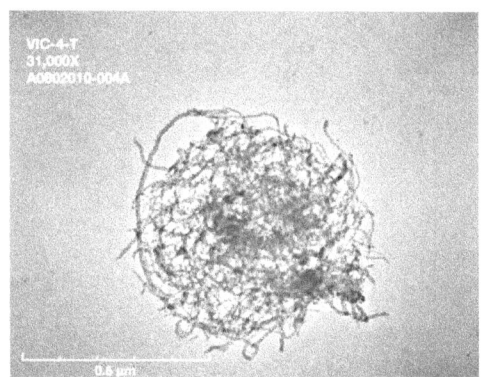

Figure 3. Aerosol droplets containing nanomaterials ejected from vial during sonication

for valuable information on process change or chemical substitution is the Toxics Use Reduction Institute at the University of Massachusetts Lowell [www.turi.org].

It is also possible to substitute a less "energetic" operating condition, and thereby modify a process to make it inherently safer. An example of process modification was demonstrated in a laboratory producing CNTs by chemical vapor deposition. Optimizing the furnace reaction temperature maximized the production of CNTs while minimizing the release of CNTs in the furnace exhaust [Tsai et al. 2009b].

6.2 Isolation and Engineering Controls

Isolation includes the physical isolation of a process or piece of equipment either by locating it in an area separate from the worker or by placing it within an enclosure that will contain the nanomaterials released. Engineering controls include any physical change to the process or workplace that reduces contaminant emissions and subsequent employee exposure. Several factors will influence the selection of exposure controls for nanomaterials, including quantity of nanomaterial handled or produced, physical form, and task duration. As each one of these variables increases, exposure risk becomes greater, as does the need for more efficient exposure control measures (Figure 4, adopted from NIOSH [2009a]). Operations involving easily dispersed dry nanomaterials deserve more attention and more stringent controls (e.g., enclosure) than those where the nanomaterials are suspended in a liquid matrix or imbedded in a solid. Liquid nanoparticle suspensions rarely pose a danger of inhalation exposure during routine operations, but they may represent a

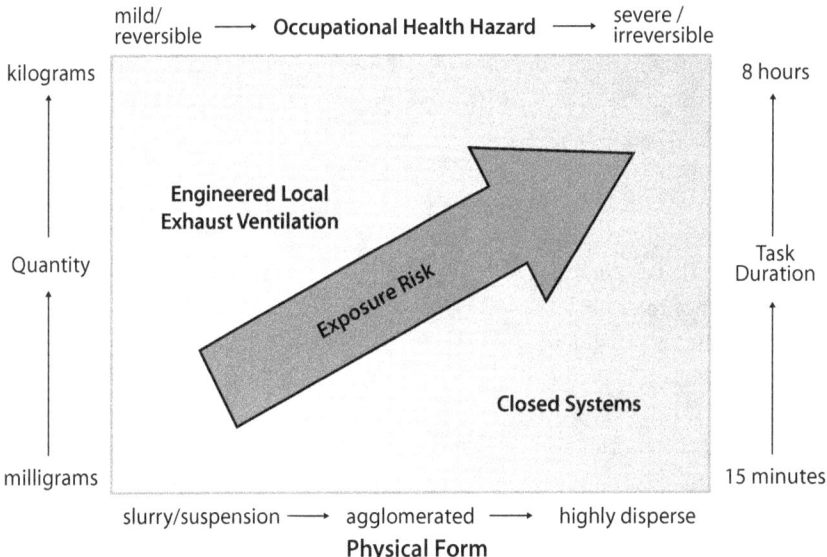

Figure 4. Factors influencing control selection

significant hazard when aerosolized or in unexpected situations such as a spill. Nanomaterials incorporated into bulk solids may pose some risk if the solid matrix is cut, sawed, drilled, sanded, or handled in any way that creates a dust or releases the nanomaterial.

6.2.1 Containment

Containment refers to the physical isolation of a process or a piece of equipment to prevent the release of the hazardous material into the workplace. An example of process isolation would be the location of a twin-screw extruder used to make CNT composites in a room separated from the rest of the research facility. An example in chemistry labs is the use of specially designed separate storage cabinets for flammables, acids, and bases. Another example of containment would be a glovebox, which is a sealed container with attached gloves that allows the researcher to carry out process or tasks while being physically separated from the hazard.

6.2.2 Ventilation

General exhaust ventilation (GEV)

It is important that any laboratory working with nanomaterials have sufficient general exhaust ventilation (GEV); however, it should not be the sole means of controlling nanomaterial exposure. GEV is typically provided by the building's heating, ventilation, and air conditioning (HVAC) system. Recommended ventilation rates for general laboratory use range from 4 to 12 air changes per hour, if LEV systems are used as the primary means of exposure control [OSHA 1990]. Laboratories should have nonrecirculating ventilation systems (preferably, 100% exhaust air), and lab pressurization should be negative to the hallway [DiBerardinis 1993]. Additionally, the air supply and air exhaust should be carefully located so that supplied air passes through the area that is being controlled. The exhaust should be as close as possible to the source of contamination, and the workers should be positioned between the air supply and the source. Exhausted air should be discharged away from windows, other air intakes, or other means of re-entry [ACGIH 2007].

Care must be taken to prevent the migration of nanomaterials into adjacent rooms or areas through the building's HVAC system, because of area pressurizations and directional airflows, or as a result of equipment and personnel moving from one area to another.

Local exhaust ventilation

A local exhaust ventilation (LEV) system with air cleaner is shown in Figure 5. Laboratory settings would have chemical fume hoods, vented enclosures, and special devices connected. The exhaust hood typically is next to or encloses the contaminant source to control exposures at the source. Air flowing into the hood entrains the contaminants and carries them through the duct, where they are either removed by an air cleaner or vented to the atmosphere. Other LEV systems include biological safety cabinets and powder-handling enclosures. Section 7.0 is dedicated entirely to LEV and its use with nanomaterials.

Use good housekeeping in laboratories where nanomaterials are handled.

- Clean all working surfaces potentially contaminated with nanomaterials (e.g., benches, glassware, apparatus, exhaust hoods, support equipment) at the end of each day with a HEPA vacuum and/or wet wiping. Do not dry sweep or use compressed air.
- As an alternative to HEPA-vacuuming lab bench tops, bench top protective covering material may be used.
- Make use of hand-washing facilities before eating, smoking, or leaving the worksite.
- Use facilities for showering and change clothes to prevent the inadvertent cross-contamination of other areas (including take-home).
- Provide laundry service for contaminated work clothing.
- Do not eat or drink in the areas where nanomaterials are handled.
- Collect laboratory waste in sealed, labeled containers approved for the particular waste stream in a manner that minimizes potential exposure during the transfer of waste into the container. Store the container in secondary containment.

6.3 Administrative Controls

Administrative controls contribute to worker exposure reduction, but they do not always reduce the airborne concentration of the contaminant in the workplace. They often include limiting exposure by reducing the time the employee is handling the material, specifying good housekeeping and other good work practices, training employees, and implementing proper labeling and storage of materials. Administrative controls in some research laboratories may include maintaining clean room conditions [Schulte et al. 2008].

Hand-washing facilities should always be used before eating, drinking, smoking, or leaving the workplace. Food and drink should not be permitted in the areas where nanomaterials are handled.

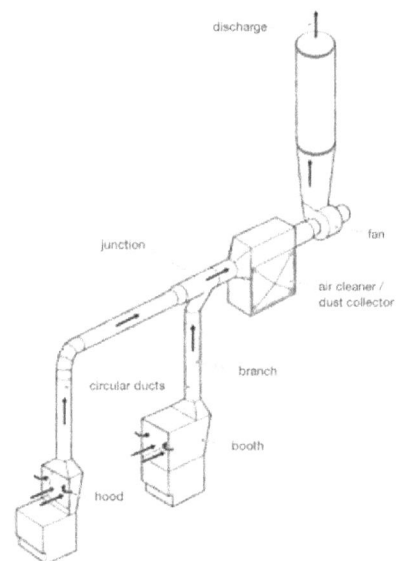

Figure 5. A local exhaust ventilation system with air cleaner

6.3.1 Employee Training

All employees working with engineered nanomaterials should receive training on the associated hazards and risks. The OSHA Hazard Communication Standard, 29 CFR

> **When working with nanomaterials, use space that is isolated as much as possible from the rest of the lab, with as few people in that space as possible.**
>
> - Keep laboratory doors closed and limit access to the laboratory (e.g., via key cards) to prevent unauthorized access.
> - Post appropriate warnings in laboratories, including measures to be taken to protect laboratory researchers and visitors from exposure risks.
> - Use local exhaust in all areas of material collection and transfer where possible.
> - Cover all containers when not in use.
> - Use 100% fresh supply air. Do not recirculate room air.

1910.1200, requires that at a minimum, training should address means to detect the chemicals in the workplace, the hazards associated with those chemicals, and procedures to prevent exposure [OSHA 1994]. In addition training should include appropriate nanomaterial handling and storage procedures, proper use of PPE, cleaning of contaminated surfaces or clothing, and proper disposal of nanomaterials or nanomaterial-contaminated objects [NIOSH 2008]. Employees should be educated regarding the job tasks that may expose them to nanomaterials and the use of appropriate controls and work practices to minimize exposure.

6.3.2 Labeling and Storage

Under the OSHA Hazard Communication Standard, 29 CFR 1910.1200, employers are required to label all hazardous chemicals in the workplace. Nanomaterials should be stored in labeled containers that indicate their chemical content and form. Liquids or dry particles should always be stored in unbreakable, tightly sealed containers. Secondary containment should be used when appropriate. Appropriate signage indicating the hazard, PPE requirements, and any other pertinent information should be posted at entry points to areas where nanomaterials and other hazardous compounds are handled or stored.

6.4 Personal Protective Equipment

Personal protective equipment (PPE) should be required when engineering and/or administrative controls are not feasible or effective in reducing exposures to acceptable levels and wherever it is necessary because of hazards. Protective equipment must be used and maintained in a sanitary and reliable condition [OSHA 2008]. Based on the uncertainty of the health risk of nanomaterials, it may be prudent to wear appropriate PPE on a precautionary basis. PPE can include respirators, gloves, clothing, face shields, safety glasses, and other garments designed to protect the wearer.

6.4.1 Protective Clothing

There are no standards or guidelines for the use or selection of protective clothing or other apparel for working with nanomaterials [OSHA 2008]. Suggested PPE consistent with basic industrial hygiene practice includes the following:

- Clothing appropriate for a wet-chemistry laboratory, including closed-toe shoes made of a low permeability material. (Disposable, over-the-shoe booties may be necessary to prevent tracking nanomaterials from the laboratory).

- Long pants (without cuffs) and a long-sleeved shirt.

- Impervious laboratory coats (noncotton). (If nondisposable laboratory coats are used, they should remain in the laboratory/change-out area to prevent nanoparticles from being transported into common areas).

- All re-useable protective clothing should be laundered. The clothing should be placed in closed bags before being taken out of the laboratory for cleaning in a central, approved location.

- Safety glasses/goggles and/or face shields as appropriate, as determined in an assessment of the hazard risk. A face shield alone is not sufficient protection against unbound dry materials.

- Nitrile or other chemically impervious gloves, as appropriate for handling nanomaterial powders and liquids. Suggested guidelines for the selection and use of gloves are as follows:

  — The proper selection of gloves should take into account the resistance of the glove to the nanomaterial (if available) and, if the nanomaterial is suspended in liquids, the liquid.

  — Chemically resistant gloves can develop cracks when they are used, so gloves should be changed whenever they show visible signs of wear.

  — Contaminated gloves should be kept in a closed plastic bag in the work area until disposal.

  — If protective clothing and/or gloves are required, particular attention should be given to preventing exposure to skin, especially abraded or lacerated skin. (Figure 6 demonstrates improper and proper use of sleeves).

  — Special attention should be given to the proper removal and disposal of contaminated gloves to prevent skin contamination.

  — Gloves should also be routinely replaced to minimize the risk of exposure and contamination of the work environment.

 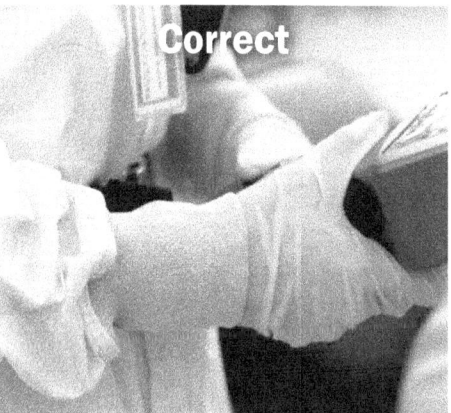

Figure 6. Make sure that the coveralls and gloves do not allow nanomaterials to contact the skin.

6.4.2 Respirators

When the potential exists that workers may inhale nanomaterials due to a lack of effective engineering controls or during activities with higher nanomaterials exposure potential (e.g., emergencies), appropriate respirators, selected according to the NIOSH Respirator Selection Logic [NIOSH 2005], should be used pursuant to the Occupational Safety and Health Administration (OSHA) respiratory protection standard 29 CFR 1910.134 [OSHA 1992]. Figure 7 shows an example of a process where a respirator was worn when working with nanomaterials.

The OSHA respiratory protection standard requires the development of a written respiratory protection program for any workplace where respirators are necessary to protect the health of the worker or whenever required by the employer. The program should include the following elements [OSHA 1992]:

- Procedures for selecting respirators for use in the workplace.
- Medical evaluations and fit-testing of employees required to use respirators.
- Procedures and schedules for cleaning, disinfecting, storing, inspecting, repairing, discarding, and otherwise maintaining the respirator.

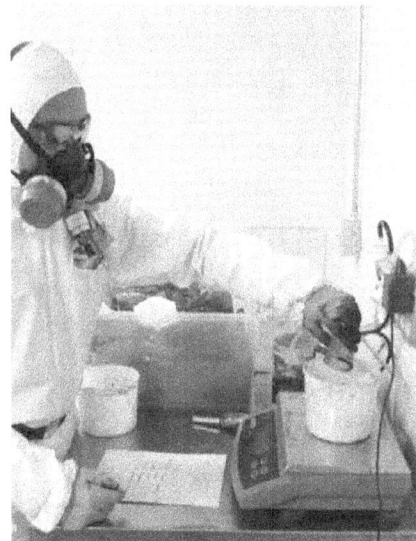

Figure 7. Weighing of carbon nanotubes.

Personal protective equipment should be used when there is the potential for exposure to the nanomaterial.

- At a minimum, for all laboratory activities, wear impervious (non-woven) laboratory coats (or coveralls, or a work uniform that covers the arms); long pants without cuffs: a long-sleeved shirt; closed-toe shoes made of a low-permeability material, or disposable foot covers; eye protection; and appropriate chemical-resistant gloves (depending on the chemical exposure).
- Respiratory and face protection and jumpsuits or chemically resistant protective clothing may be needed for laboratory activities, depending on the hazard or quantities of the material(s) handled, the availability of appropriate controls, and the exposure risks.
- Respiratory protection should be selected in consultation with the laboratory occupational safety and health professional according to the NIOSH Respirator Selection Logic.
- Follow proper procedures to properly select, maintain, don, doff, and decontaminate personal protective clothing and equipment as described in the laboratory risk management plant.
- Provide an area outside of the contaminated area for donning and doffing PPE.

- Training of employees in the respiratory hazards to which they may be exposed, and proper use and maintenance of the respirator.
- Procedures to evaluate the effectiveness of the program.

The NIOSH proposed recommended exposure limit (REL) of 7 µg/m^3 elemental carbon as an 8-hr Time Weighted Average respirable mass airborne concentration for carbon nanotubes and carbon nanofibers, and 0.3 mg/m^3 for ultrafine titanium dioxide suggest respirators may be necessary if expected exposures are above this level [NIOSH 2010a, 1994, 2011]. A properly fit-tested, half-face particulate respirator will provide protection at exposure concentrations 10 times the REL, while an elastomeric full facepiece respirator with P100 filters will provide protection at 50 times the REL. NIOSH provides further guidance for selecting respirators in the NIOSH Respirator Selection Logic 2004 [NIOSH 2005].

7 Local Exhaust Ventilation

Local Exhaust Ventilation (LEV) systems reduce or prevent exposure to airborne contaminants by capturing them at their source. The first element of a LEV system is the hood, of which there are two basic types: enclosing and exterior [ACGIH 2007]. The most common LEV system used in research laboratories is the laboratory chemical hood. An enclosing laboratory chemical hood is shown in Figure 8. The hood, with its moving sash, is actually only a partial, "three-sided" enclosure. Other common types of LEV include the exterior hood (Figure 9), which is placed adjacent to the contaminant source. Due to the unknown hazard potential of nanomaterials, a more conservative approach to ventilation control is dictated, with an emphasis on enclosing systems.

In general, enclosing hoods are preferred to exterior hoods, because the contaminants are contained inside the hood itself. The hood provides a barrier between the worker and the contaminant (this barrier is only partial for a laboratory chemical hood when the sash is opened—see the discussion on these hoods below). Sufficient airflow must be provided through any openings in the enclosure to ensure that the contaminants don't escape the hood. Airflow through openings is usually specified as a certain required face velocity at the opening, and the value depends on the hood design and application [ACGIH 2007]. It can be difficult to choose the proper face velocity to achieve adequate flow for enclosing hoods with large openings. Most laboratory chemical hoods are designed to operate at 100 ft/min face velocity, which can create problems because of turbulence. Turbulence may reduce the capture efficiency of the hood and may disturb settled particles [NRC 2011]. This is discussed more fully in the following section. Exterior hoods are less preferred because they must create a capture velocity at the point of contaminant generation to capture the contaminant and draw it into the hood.

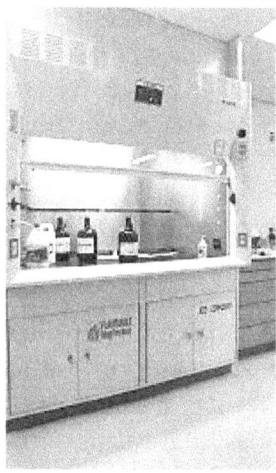

Figure 8. Laboratory chemical hood (note that the work area is *inside* the hood)

7.1 Laboratory Chemical Hoods

A properly designed and maintained chemical fume hood can offer significant worker protection if used properly. There are many different hood designs, but the most common categories are the conventional or constant-flow hood, the bypass hood, and the variable air volume constant-velocity hood. Examples of each are shown in Figures 10–12. Positive pressure laminar flow hoods that are designed for product protection and direct potentially contaminated air toward the user are not considered local exhaust ventilation and are not recommended for use to protect from nanomaterial exposures [Harford 2007]. All chemical hoods have certain common design elements, including an exhaust fan to move air through the hood, a moving sash, exhaust slots, and a horizontal work surface. The sash can move in either a vertical or a horizontal direction. A crucial performance element for any chemical hood is the face velocity, defined as the

Figure 9. Exterior hood (note that the work area is *in front* of the hood entrance)

average air velocity at the face of the hood at the sash opening. Maintaining a constant, minimum face velocity provides confidence that operations (and hazardous agents) within the hood will be contained. Hood face velocity must be evaluated and controlled by the facility's engineering or health and safety staff. The current consensus of the literature is that the average face velocity for a laboratory chemical hood should be in the range of 80–120 ft/min [Burgess et al. 2004]. The flow control system on a constant-velocity hood should be adjusted to give a face velocity in this range. Each chemical hood should be clearly marked with the proper hood sash location that will give the desired face velocity; depending on the hood design, this could be a single location or a range of locations. Containment verification using tracer gases to provide quantitative data and smoke testing to visual airflow patterns is recommended when the hood is installed, when substantial changes are made to the ventilation system, and periodically as part of a preventive maintenance program. Testing should be performed following the ANSI/ASHRAE 110 or equivalent protocol [NRC 2011].

In addition to the face velocity, it is important that the airflow be distributed evenly across the hood face. ANSI/AIHA Z9.5 recommends that variations of face velocity across the hood face should be within ±20% of the average face velocity; however, some laboratories select a stricter standard of ±10%.

The constant-flow hood (Figure 10) constitutes the oldest, simplest chemical hood design. The exhaust fan introduces a constant volumetric airflow moving through the sash opening. For this hood design, the face velocity is lowest when the sash is wide open; when the sash is lowered the face velocity increases.

The bypass hood (Figure 11) maintains a constant hood face velocity and incorporates a bypass grille located above the sash opening. When the sash is wide open it blocks the bypass grille, allowing all of the air to flow through the hood opening. As the sash is lowered, it uncovers increasingly greater amounts of the bypass grille, allowing increasing amounts of air to flow through this alternative path. If it is designed and operated properly, the amount of air flowing through the bypass grille is just sufficient to maintain a constant face velocity. Typically, however, this constant velocity can be maintained over a certain part of the sash's total range.

The constant-velocity hood (Figure 12) uses a control system to detect the sash position, face velocity and system pressure, and change the fan motor speed or other mechanism, such as mechanical dampers, to increase the airflow when the sash is raised and decrease it when the sash is lowered, thus maintaining a constant face velocity.

Tsai et al. [2010] evaluated the efficiency of these three main types of laboratory chemical hoods to reduce exposure to aluminum oxide nanomaterials while manually handling them inside the chemical hoods. They determined that the particle release to the

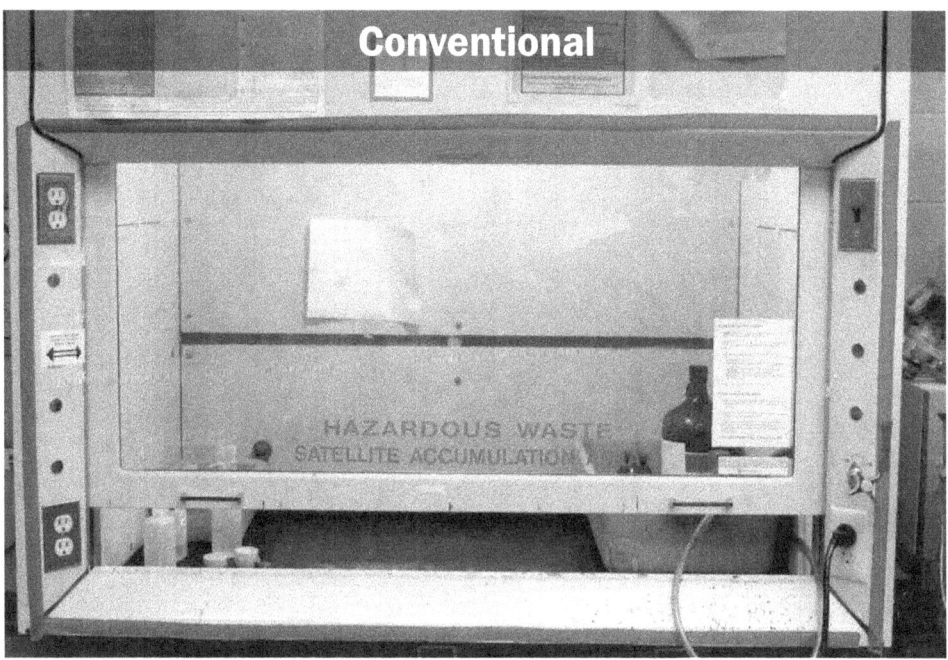

Figure 10. Conventional (constant-flow) laboratory chemical hood.

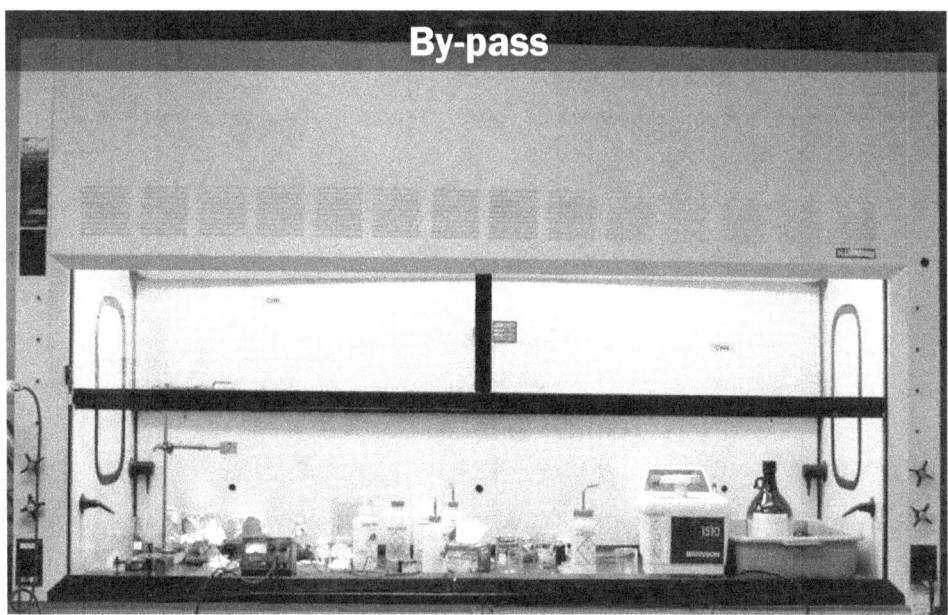

Figure 11. Laboratory bypass hood (note the by-pass chamber above the sash).

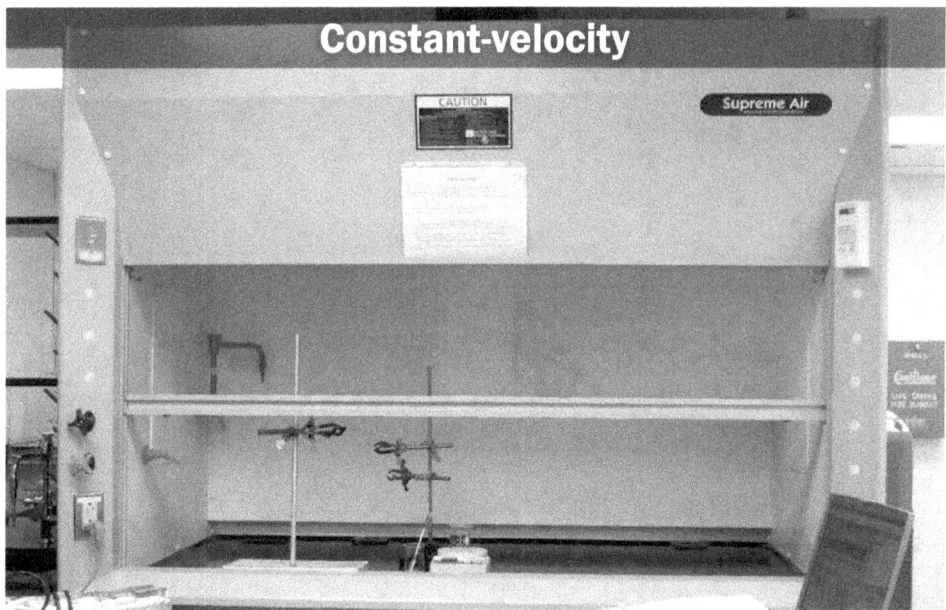

Figure 12. Laboratory constant-velocity hood (note the flow monitor on the sash column)

worker's breathing zone was greatest when using a constant-flow hood, as compared to a bypass and a constant-velocity hood.

Saunders [1993] and DiBerardinis [1993] described basic elements that all chemical hoods should incorporate. All laboratory chemical hoods should include, *at a minimum*, the following design elements:

- A minimum width of 4 feet (wider is better, to allow flexibility in equipment use).
- A minimum sash open height of 30 inches.
- A bottom-front airfoil.
- A sash that is easily movable over its entire range of motion.
- A sash that holds its position over its entire range of motion.
- Side walls that are smooth, rounded, and tapered toward the inside of the sash opening [Schulte et al. 1954].

In addition, the following factors relative to the hood location are very important for proper hood performance

- Air currents outside a hood may disrupt the airflow at the face and therefore impact the ability of a hood to contain the contaminant.

- The hood should not be located next to any laboratory entry door or any other high-traffic location.
- The hood should be at least 5 feet from any HVAC air supply grille; a distance of 10 feet is preferred.

The following practices are important for working in laboratory chemical hoods:

- The hood sash should be kept wide open during equipment set-up only; during actual use, the sash should be lowered to the position that gives proper hood face velocity.
- Equipment should be at least 6 inches behind the sash opening (many hoods have a recessed floor starting at this distance, to encourage proper use).
- When working in the hood, the user should avoid working at the edge of the hood and should minimize arm movements; all such movements should be slow and smooth.
- Traffic past the hood should be minimized when nanomaterial powders are being manipulated. Research has shown that the passage of a person past the hood face at walking speeds creates a turbulent wake sufficient to pull contaminants from the hood [Johnson and Fletcher 1996].
- During experiments, when no access is required, the sash should be kept either in the same position as when work is performed (constant flow and bypass hoods) or lowered to the fully closed position (constant velocity hoods).
- When using a local exhaust system, do not directly exhaust into the work environment any effluent (air) that is reasonably suspected to contain nanomaterials. The exhaust air should be passed through a HEPA filter [NIOSH 2007] and, when feasible, released outside the facility. If the exhausted air is recirculated, then steps should be taken to ensure that recirculated air doesn't contain the engineered nanoparticle.
- Handle exhaust filters from the chemical hoods in a manner that minimizes exposure. Put a plastic-lined bag around the filter at the source when removing it so that particulates are not potentially released to the work environment. Wear appropriate PPE during all maintenance and cleaning activities.
- Storage of materials in the chemical hood should be minimized or eliminated. Materials stored in the hood can adversely affect the containment by disrupting airflow. If items must be placed inside the hood, make sure they are placed near the back and do not block the air slots.

7.1.1 Working with Nanomaterial Powders in Chemical Hoods

Research performed at the University of Massachusetts Lowell [Tsai et al. 2009a] has demonstrated that nanomaterial powders may be released back into the work area from chemical fume hoods during tasks such as weighing or transferring from container to container. Releases that are not detectable on a mass basis were found to have a very high particle number concentration. Experiments performed on constant-volume and bypass hoods demonstrated that working with the sash either too low or too high could cause nanoparticles to escape from the hood. When the sash is too high, the face velocity can fall below the recommended minimum of 80 ft/min. This low face velocity and the large opening created by the high sash allow random room air currents to enter the hood, entrain airborne nanomaterials, and carry them out of the hood. When the sash is too low, the face velocity can exceed the recommended maximum of 120 ft/min. This causes a strong turbulent wake in the space between the worker and the hood face, which can pull airborne nanomaterials from the hood. Because of the possibility of loss of the nanomaterial at high face velocities, the correct sash height should be determined for the specific process being carried out, based on the ability of the chemical hood to capture the nanomaterial. Because of the potential to create turbulence, the hood should be as uncluttered as possible, and the researcher should remove his arms or other objects from the hood very slowly [Tsai et al. 2009a]. If the potential for material loss exists or if exhaust filtration is infeasible, alternative exhausted enclosures should be considered such as low flow enclosures or biological safety cabinets (see section 7.2).

7.1.2 New Hood Designs

Researchers are designing new lower-flow chemical hoods that may offer improved performance for handling nanoparticle powders. In some studies, it has been noted that hood face velocities of 100 ft/min may result in the loss of nanomaterial [Johnson et al. 2010]. Lower-flow chemical hoods operate with face velocities of less than 100 ft/min. However, at this time, there is very little research on the effectiveness of low-flow fume hoods for handling nanoparticle powders. A recent hood design approach is the air-curtain hood [Huang et al. 2007], which uses a downward air jet emanating from a double-pane sash to isolate the interior of the hood from the exterior environment. An evaluation of the hood at the University of Massachusetts Lowell [Tsai et al. 2010] indicated that it can be effective at containing airborne nanoparticles.

7.2 Alternatives to Conventional Chemical Hoods

7.2.1 Glovebox Enclosures

A higher level of protection for handling dry powders is obtained by using a glovebox enclosure [DiBerardinis 1993] (Figure 13). The primary advantage of using a glovebox is the protection it affords; when used properly to manipulate nanoparticle powders, a glovebox should prevent exposure to the user. The disadvantages of using a glovebox relate to the extra time required to move materials and equipment in and out of the enclosure, the difficulty of manipulating nanomaterials when wearing gloves, and the need to periodically clean the enclosure. The two most likely sources of exposure when using a glovebox are the transfer of materials into and out of the box and the cleaning of the box following its use. Both of these activities must be performed with extreme care. Note that glovebox enclosures are sometimes used under positive pressure with respect to the surrounding room (e.g., as shown in Figure 13), with an inert atmosphere such as nitrogen to reduce the risks of fire, explosion, or oxidation. Such use can increase the possibility of airborne releases from the enclosure. Proper leak-testing procedures, in accordance with the American Glovebox Society Standards, should be followed to verify containment.

7.2.2 Biological Safety Cabinets

Biological safety cabinets (BSCs) serve as a primary means of containment developed for working safely with infectious microorganisms, such as viruses, bacteria, and fungal spores [Chosewood et al. 2009]. BSCs are designed to provide personnel, environment,

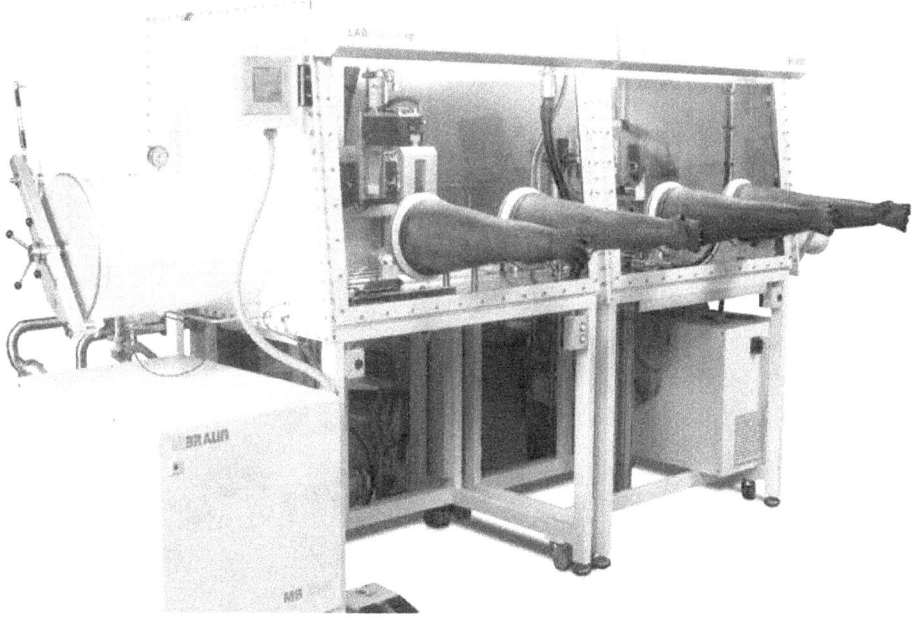

Figure 13. Glove box enclosure (shown here under positive pressure for use with inert atmospheres, rather than for enhanced containment of particles and gases)(MBRAUN; used with permission).

and product protection when appropriate practices and procedures are followed. Nanomaterials, whose size range is similar to that of bioaerosols (microorganisms that are suspended in the air), should behave aerodynamically in the same manner. Additionally, the HEPA filtration systems in BSCs should be equally effective in filtering nanomaterials and bioaerosols because of primary particle size. Therefore, it is reasonable to assume that these cabinets will offer similar levels of protection against bioaerosols and airborne nanomaterials. Three different classes of biological safety cabinets are defined as follows:

- A Class I biological safety cabinet resembles a chemical hood, with the additional requirement that the exhaust air must be treated before it is discharged to the atmosphere.

- A Class II biological safety cabinet is designed to protect the operator, the product, and the environment (Figure 14). It has an inward airflow through the open sash to protect the operator, a downward flow of HEPA-filtered air to protect the product, and a HEPA-filtered exhaust to protect the environment. Class II cabinets are designed for use against low- to moderate-risk biological agents. The four types of Class II cabinets are defined as A, B1, B2, and B3; each type of Class II cabinet has different air recirculation percentages, and the level of control increases from A to B3. Because air is recirculated in Class II type A and B1 cabinets, tasks involving volatile materials should not be performed in these cabinets.

- A Class III biological safety cabinet is a highly sophisticated glovebox. The sealed enclosure is maintained at a negative static pressure of at least 0.5 inches H_2O, the supply air is HEPA-filtered, and the exhaust air is either double-HEPA-filtered or passed through a single HEPA filter and then incinerated. Class III cabinets are meant for the highest-risk biological agents.

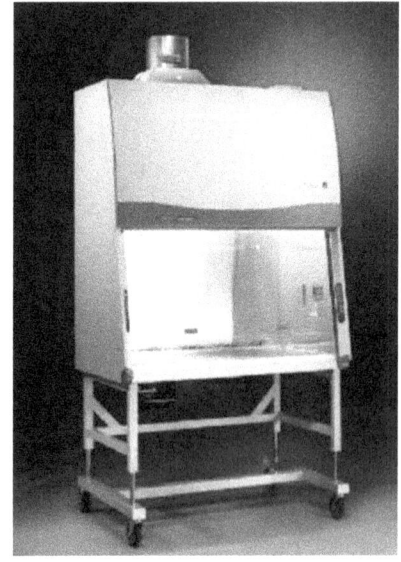

Figure 14. Class II Biological Safety Cabinet (Labconco Corporation; used with permission)

The most widely used class of biological safety cabinet is Class II; this is the class most likely to be available to researchers working with nanomaterials. Because this cabinet type has an inward airflow through the sash, similar to a laboratory chemical hood, it may be appropriate for use against dry powder chemicals as well as biological agents. Caution must be taken, however, because the complex airflow patterns inside a Class II cabinet create complex turbulence patterns that may adversely affect the researcher's ability to handle nanomaterials without loss. Class II, type A1 and A2 cabinets should be exhausted outside of the building via a "thimble" connection to avoid disturbing the internal cabinet airflow, whereas type B1 and B2 cabinets should be hard-ducted [Chosewood 2009]. Researchers handling biological hazards in addition to nanomaterials should follow all applicable regulations.

7.2.3 Powder Handling Enclosures

For a number of years, equipment manufacturers have offered ventilated enclosures specifically for weighing and manipulating small quantities of dry powders. These were first developed and marketed to the pharmaceutical industry, but they are now sold as general purpose powder-handling enclosures. Systems can be self-contained, with their own fan and HEPA filtration unit (an example is shown in Figures 15 and 16), or connected to a central exhaust system. The exhaust can be ducted to the outside or

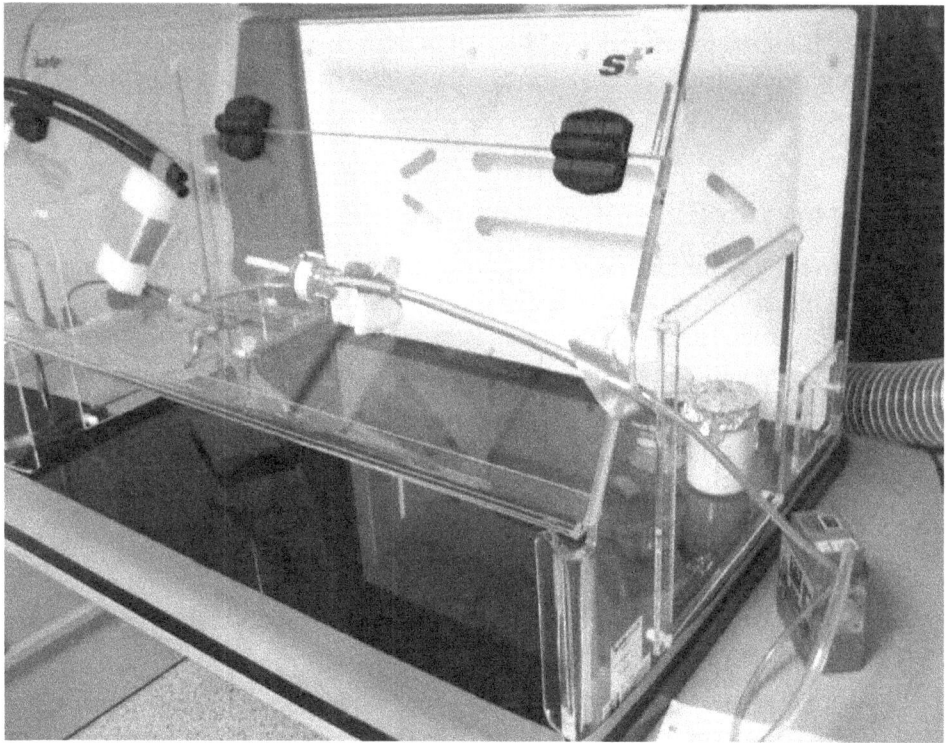

Figure 15. Powder-handling enclosure

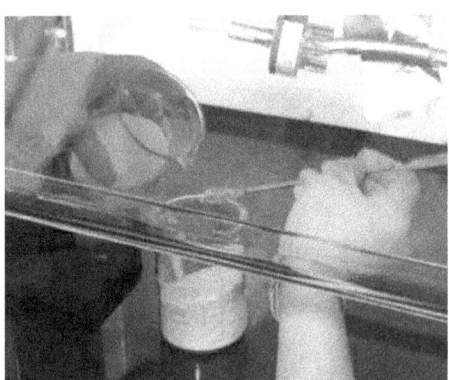

Figure 16. Close-up of a nanopowder transfer operation.

Figure 17. Example of a U-Frame antistatic device being used with a microbalance (photo courtesy of METTLER TOLEDO).

recirculated into the room. One advantage of these devices is that they operate at much lower flow rates and velocities than the chemical hoods. The internal turbulence is reduced significantly, lessening the potential for loss or ejection of the nanomaterial.

Figure 17 illustrates an electrostatic discharge unit that can be used to reduce electrostatic charge on nanomaterials prior to transferring them from one container to another or to a weighing station.

8 Methods for Exposure Control Verification

When verifying the effectiveness of exposure control measures, it is generally preferred to measure the agent of interest using an exposure metric that directly relates to its toxicological properties. However, for most nanomaterials, sufficient data are not available to determine the most appropriate exposure metric. Effectiveness of controls can be verified by the following means:

- Testing and certification procedures specified by ANSI Z9.5 and in ASHRAE 110.
- Qualitative indicators of proper installation and functionality of the control systems (e.g., are gaskets, shrouds, and ventilation hoses in their required locations and free of visible defects?).
- Quantitative indicators of proper installation and functionality of the control systems (e.g., hood face velocities within proper ranges).
- Semi-quantitative measures of potential worker exposures, such as determinations of airborne dust concentrations (e.g., airborne particle concentrations) near the exposure control device (e.g., near the LEV, at the opening of the chemical hood).

- Quantitative measures of worker exposures (e.g., personal sampling for the nanomaterial of interest).

Verification has two purposes: (1) to ensure that the mechanical and procedural aspects of the implemented controls are performing as specified, designed, and installed; and (2) to ensure that implemented controls are maintaining nanomaterial concentrations at or below the preset limit.

Verification is essential for the following reasons:

- Factors such as area pressurization, directional airflow, dilution ventilation rates, and filtration efficiency can change.
- General or individual work practices can change.
- Task frequency and duration can change.

The verification begins with prioritization of all operations in which exposures may occur and selection of those processes in which samples will actually be taken, on the basis of professional judgment. This enables an appropriate and effective focus of resources.

As noted in the NIOSH *Approaches to Safe Nanotechnology: Managing the Health and Safety Concerns Associated with Engineered Nanomaterials* [NIOSH 2009a], exposure assessment and control verification approaches can be performed with traditional industrial hygiene sampling methods that include the use of samplers placed at static locations (area sampling), samples collected in the breathing zone of the employee (personal sampling), or measurements with real-time devices. The assessment should use both particle counters and filter-based samples [NIOSH 2009a]. Filter-based samples can be used to identify the nanomaterial of interest with electron microscopy and elemental analysis (Figure 18).

In general, personal sampling is preferred to ensure an accurate representation of the worker's exposure, whereas area samples (e.g., size-fractionated aerosol samples) and real-time (direct-reading) exposure measurements may be more useful for evaluating the need for improvement of engineering controls and work practices. Other sampling techniques can be used to measure airborne nanomaterials, but they require more expertise in their use and interpretation of the data. Selected use of these advanced methods can produce useful data for evaluating occupational exposures with respect to particle size, surface area, and morphology.

9 Periodic Re-evaluations of the Risk Management Program

Re-evaluations of the risk management program should be conducted on a scheduled periodic basis (e.g., annually) and when new information becomes available or changes occur in the workplace. Re-evaluations can foster iterations among the hazard tasks and control steps to optimize application of the hierarchy of control.

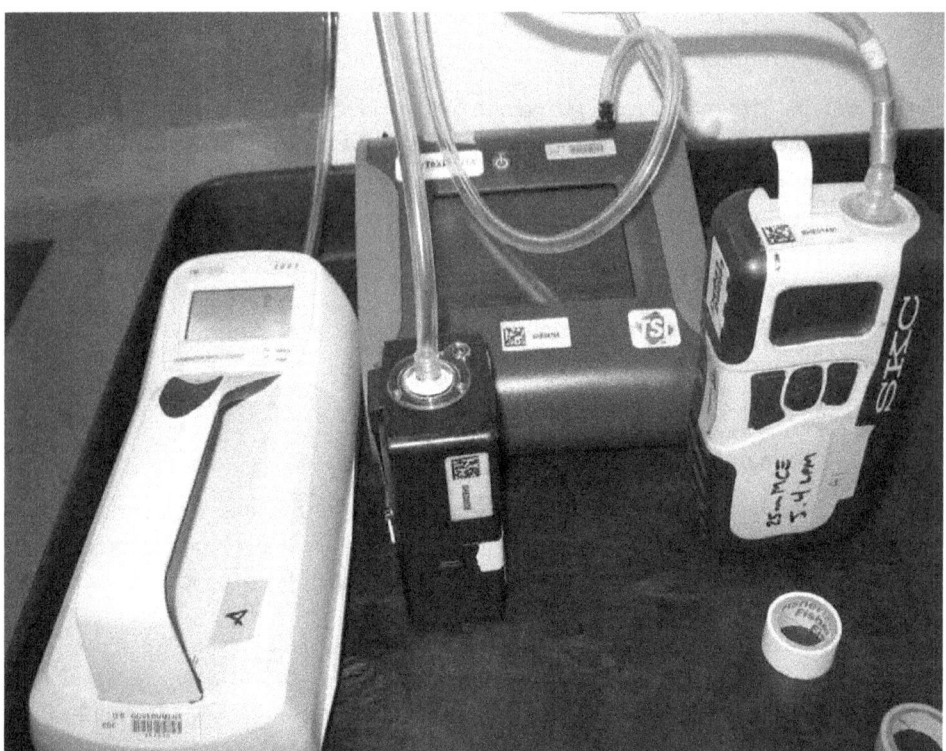

Figure 18. Area sampling for airborne nanomaterials

It is prudent to re-evaluate the risk management program when the following occur:

- Process or material modifications are made.
- New materials are introduced into the workplace.
- Modifications are made to the flow of work.
- Tasks are moved to a new location or workforce.
- New equipment is designed or installed.
- Production volume, speed, or frequency changes.
- Duration changes for operations with identified exposure risks.
- A new nanomaterial is handled.
- Physical form changes (for example, powders rather than suspensions).
- New equipment is designed or installed.
- New toxicology data are obtained.
- Medical surveillance trending suggests adverse effects.

- Occupational illness is reported.
- The workforce changes.
- A validated sampling and analytical method is developed for the nanomaterial(s) being used.
- Qualitative indicators (such as odor, visual observations, or employee reports) or quantitative indicators (such as measurements) of exposure suggest a change in control effectiveness.

10 Guidance on Developing a Control Scheme (Control Banding)

Control banding is a qualitative strategy for assessing and managing hazards associated with chemical exposures in the workplace. The concept is used to manage exposures to potentially hazardous materials through the application of one of four recommended control approaches. This concept is based on the premise that although many chemical hazards exist, there are a limited number of controls available. To determine the appropriate control strategy, one must consider the characteristics of the substance, the potential for exposure, and the hazard associated with the substance. As the potential for harm to the worker increases, so does the degree of control needed to manage the risk [NIOSH 2009c].

The four control bands are usually the following:

Band 1: Use good industrial hygiene practice and general ventilation.
Band 2: Use an engineering control, typically local exhaust ventilation.
Band 3: Enclose the process.
Band 4: Seek expert advice.

There are several control banding tools developed for use with nanomaterial exposures, [Paik et al. 2008; Zalk et al. 2009; *GoodNanoGuide* 2009; Safe Work Australia 2012]. The *GoodNanoGuide* (www.goodnanoguide.org) is an Internet-based platform for the exchange of ideas on handling nanomaterials, and it recommends a simplified approach to control banding of nanomaterials (Figure 19). With this approach, nanomaterials are grouped into three hazard groups: (A) known to be inert, (B) understand reactivity and function, or (C) unknown properties. The exposure duration is described as Short (<4 hours/day, 2 days/week), Medium (4–6 hours/day, 3–5 days/week) or Long (>6 hours/day, 3–5 days/week). The potential for exposure is described through the state of the nanomaterial: bound (nanoparticles in a solid matrix), potential release (nanoparticles in friable matrix), or free/unbound (nanoparticles unbound, not aggregated). These elements are used to determine the recommended control band.

Exposure Duration	Bound Materials	Potential Release	Free / Unbound
Hazard Group A (Known to be inert)			
Shout	1	1	2
Medium	1	1	2
Long	1	2	2
Hazard Group B (Understand reactivity/function)			
Short	1	2	2
Medium	1	2	3
Long	1	3	3
Hazard Group C (Unknown Properties)			
Short	2	2	3
Medium	2	3	4
Long	2	4	4

☐ Band 1: Use good industrial hygiene practice and general ventilation.
☐ Band 2: Use an engineering control, typically local exhaust ventilation.
☐ Band 3: Enclose the process.
■ Band 4: Seek expert advice.

Figure 19. GoodNanoGuide control banding matrix

Another tool, the CB Nanotool, bases the control band for a particular task on the overall risk level (RL), which is determined by a "severity" score and a "probability" score (Figure 20). The severity score is determined by the sum of points assigned to the following factors: surface chemistry, particle shape, particle diameter, solubility, carcinogenicity, reproductive toxicity, mutagenicity, dermal toxicity, and hazard potential of the nanomaterial and the macro-parent material. The overall probability score is based on the following elements: estimated amount of nanomaterial used during the task, dustiness or mistiness, number of employees with similar exposures, frequency of operation, and duration of operation [Paik et al. 2008]. The CB Nanotool is being used at the Lawrence Livermore National Laboratory (LLNL) and can be downloaded at http://controlbanding.net/Home.html.

One limitation of the CB Nanotool and other control banding tools for nanomaterials is that there are very few toxicological data on which to recommend control levels, other than the highest two levels, and to evaluate the validity of the tool. As health hazard studies continue to expand, and the understanding of the toxicity of nanomaterials improves, the severity scores may be adjusted to reflect the new knowledge and thereby affect the severity score to elicit a more defined control band [Zalk et al. 2009].

Probability

Severity		Extremely Unlikely (0–25)	Less Likely (20–50)	Likely (51–75)	Probable (76–100)
	Very High (76–100)	RL 3	RL 3	RL 4	RL 4
	High (51–75)	RL 2	RL 2	RL 3	RL 4
	Medium (26–50)	RL 1	RL 1	RL 2	RL 3
	Low (0–25)	RL 1	RL 1	RL 1	RL 2

Control bands:

RL 1: General ventilation
RL 2: Fume hoods or local exhaust ventilation
RL 3: Containment
RL 4: Seek specialist advice

Figure 20. Risk level matrix for the CB Nanotool

The Australian Control Banding tool is specific to carbon nanotubes [Safe Work Australia 2012]. The exposure potential is based on the amounts and types of activities, and determines the control band.

11 Fire and Explosion Control

Both carbon-containing and metal dusts can explode if they are aerosolized at a high enough concentration and if oxygen and an ignition source are present. Because the surface-to-volume ratio increases as a particle becomes smaller, nanoparticles may be more prone to explosion than an equivalent mass concentration of larger particles. In general, the potential and severity of nanomaterial explosions increase proportionally to the quantity of combustible nanomaterials being used. Thus, bench-scale research should present fewer explosion risks than work in pilot plants or full-scale manufacturing facilities. Nonetheless, all researchers should avoid creating large, highly concentrated aerosols of combustible nanomaterials.

12 Management of Nanomaterial Spills

Procedures should be developed to protect employees from exposure to nanomaterials during the cleanup of spills and spill-contaminated surfaces. Inhalation and dermal exposures will likely present the greatest risks. The potential for inhalation exposure during cleanup will be influenced by the likelihood of nanomaterials becoming airborne, with powder form presenting a greater inhalation potential than nanomaterials in solution, and liquids in turn presenting a greater potential risk than encapsulated nanomaterials.

Until relevant health and workplace exposure information is available, it is prudent to base strategies for dealing with spills and contaminated surfaces on the use of current good practices such as dust control and suppression. Standard approaches for cleaning powder spills can be used for cleaning surfaces contaminated with dry powder nanomaterials. These include access control, using HEPA-filtered vacuum cleaners, wiping up dry powders with damp cloths, or wetting the powder before wiping. Liquid spills containing nanomaterials can typically be cleaned by applying absorbent materials/liquid traps. If vacuum cleaning is employed, HEPA-filtered vacuums should be used, and care should be taken that HEPA filters are installed properly and that vacuum bags are changed according to the manufacturer's recommendations. Dry sweeping or air hoses should not be used to clean work areas. As in the case of any material spills or cleaning of contaminated surfaces, the handling and disposal should follow all applicable federal, state, and local regulations.

Equipment to contain and clean a nanomaterial spill should be readily available in or near each laboratory working with such materials. A nanomaterial spill kit for a laboratory environment may contain the following:

- Barricade tape.
- Nitrile or other chemically impervious gloves.
- Elastomeric respirator with appropriate filters.
- Adsorbent material.
- Wipes.
- Sealable plastic bags.
- Walk-off mat (e.g., Tacki-Mat®).
- HEPA-filtered vacuum.
- Spray bottle with deionized water or other appropriate liquid.

13 Occupational Health Surveillance

Occupational health surveillance involves the ongoing systematic collection, analysis, and dissemination of exposure and health data on groups of workers for the purpose of preventing illness and injury [NIOSH 2009b]. Occupational health surveillance, which includes hazard and medical surveillance, is an essential component of an effective occupational safety and health program [Harber et al. 2003; Baker and Matte 2005; NIOSH 2006; Wagner and Fine 2008]. NIOSH continues to recommend occupational health surveillance as an important part of an effective risk management program for nanomaterial workers.

Medical screening in the workplace focuses on the early detection of health outcomes for individual workers and may involve an occupational history, medical examination, and application of specific medical tests to detect the presence of toxicants or early pathologic changes before the worker would normally seek clinical care for symptomatic presentations. Medical screening and resulting interventions represent secondary prevention and should not replace primary prevention efforts to minimize employee exposures to nanomaterials. Medical surveillance involves the ongoing evaluation of the health status of a group of workers through the collection and aggregate analysis of health data for the purpose of preventing disease and evaluating the effectiveness of intervention programs.

Specific guidance for workers exposed to Carbon Nanotubes or Nanofibers is described in the Draft NIOSH *Current Intelligence Bulletin: Occupational Exposure to Carbon Nanotubes or Nanofibers* [NIOSH 2010a]. NIOSH has developed interim guidance relevant to medical screening (one component of an occupational health surveillance program) for nanotechnology workers (see NIOSH *Current Intelligence Bulletin: Interim Guidance on Medical Screening of Workers Potentially Exposed to Engineered Nanoparticles* [http://www.cdc.gov/niosh/docs/2009-116]).

If medical screening recommendations exist for chemical or bulk materials of which nanomaterials are composed, they would apply to nanomaterials as well. A basic medical surveillance program should contain the following elements [Trout and Schulte 2010]:

- An initial medical evaluation performed by a qualified health professional and other examinations or medical tests deemed necessary by the health professional.
- Periodic evaluations including symptoms surveys, physical exams, or specific medical tests based on data gathered in the initial evaluation.
- Post-incident evaluations.
- Employee training.
- Periodic analysis of the medical screening data to identify trends or patterns.

14 Conclusions

Given the growing body of knowledge about the potential hazards presented by worker exposure to engineered nanomaterials, it is important to protect researchers, laboratory staff, and others who work in the laboratory (e.g., janitors). The full range of occupational hygiene controls will be necessary to limit exposures to nanomaterials as a means to prevent adverse health outcomes in the research community. Engineering and administrative controls can eliminate or minimize the amount of nanomaterials that will be present in workplace air or settled on surfaces. Personal protective equipment can be used where other types of controls are not available or practical.

Nanomaterial health and safety is a rapidly evolving field that must respond to new information regarding nanomaterial toxicity and exposure potential. Thus, it is recommended that researchers and health and safety professionals stay abreast of new developments in nanomaterial workplace protection as they are published, both in the peer-reviewed literature and on credible Web sites such as those of NIOSH [www.cdc.gov/niosh] and the GoodNanoGuide [www.goodnanoguide.org].

References

ACGIH [2007]. Industrial ventilation: A manual of recommended practice for design. 26th ed. Cincinnati, OH: American Congress of Governmental Industrial Hygienists.

Ahmad Z [2006]. Processing and synthesis techniques for the preparation of nanomaterials. Dhahran, Saudi Arabia: Mechanical Engineering Department, King Fahd University of Petroleum & Minerals [http://www.azonano.com/article.aspx?ArticleID=1710]. Date accessed: July 20, 2011.

Aitken R, Creely KS, Tran CL [2004]. Nanoparticles: an occupational hygiene review. London, UK: Health and Safety Executive [http://www.hse.gov.uk/research/rrhtm/rr274.htm]. Date accessed: July 20, 2011.

Baker EL, Matte TP [2005]. Occupational health surveillance. In: Rosenstock L, Cullen MR, Brodkin CA, Redlich CA, eds. Textbook of clinical occupational and environmental medicine. Philadelphia, PA. Elsevier Saunders Company, pp. 76–82.

Barlow PG, Clouter-Baker A, Donaldson K, Maccallum J, Stone V [2005]. Carbon black nanoparticles induce type II epithelial cells to release chemotoxins for alveolar macrophages. Part Fibre Toxicol 2:11.

Behrens I, Pena AI, Alonso MJ, Kissel T [2002]. Comparative uptake studies of bioadhesive and non-bioadhesive nanoparticles in human intestinal cell lines and rats: the effect of mucus on particle adsorption and transport. Pharm Res 19(8):1185–1193.

Brown DM, Wilson MR, MacNee W, Stone V, Donaldson K [2001]. Size-dependent proinflammatory effects of ultrafine polystyrene particles: a role for surface area and oxidative stress in the enhanced activity of ultrafines. Toxicol Appl Pharmacol 175(3):191–199.

Burgess WA, Ellenbecker MJ, Treitman RD [2004]. Ventilation for control of the work environment. 2nd ed. Hoboken, N.J.: Wiley-Interscience, pp. xvi.

Buzea C, Pacheco II, Robbie K [2007]. Nanomaterials and nanoparticles: sources and toxicity. Biointerphases 2(4):MR17–MR71.

Chattopadhyay K [1992]. Nanocomposites by rapid solidification route. Bulletin of Materials Science 15(6):515–525.

Chosewood LC, Wilson DE, eds. [2009]. Biosafety in microbiological and biomedical laboratories. 5th ed. Washington, DC: U.S. Department of Health and Human Services, Public Health Service, Centers for Disease Control and Prevention, National Institutes of Health HHS Publication No.(CDC) 2-112.

Demou E, Stark WJ, Hellweg S [2009]. Particle emission and exposure during nanoparticle synthesis in research laboratories. Ann Occup Hyg 53(8):829–838.

DiBerardinis LJ [1993]. Guidelines for laboratory design: health and safety considerations. 2nd ed. New York: Wiley, pp. xiv.

Donaldson K, Aitken R, Tran L, Stone V, Duffin R, Forrest G, Alexander A [2006]. Carbon nanotubes: a review of their properties in relation to pulmonary toxicology and workplace safety. Toxicol Sci 92(1):5–22.

Duffin R, Tran C, Clouter A, Brown D, MacNee W, Stone V, Donaldson K [2002]. The importance of surface area and specific reactivity in the acute pulmonary inflammatory response to particles. Ann Occup Hyg 46(suppl. 1):242–245.

Duffin R, Tran L, Brown D, Stone V, Donaldson K [2007]. Proinflammogenic effects of low-toxicity and metal nanoparticles in vivo and in vitro: highlighting the role of particle surface area and surface reactivity. Inhalation Toxicol 19(10):849–856.

Elder A, Gelein R, Silva V, Feikert T, Opanashuk L, Carter J, Potter R, Maynard A, Ito Y, Finkelstein J, Oberdörster G [2006]. Translocation of inhaled ultrafine manganese oxide particles to the central nervous system. Environ Health Perspect 114(8):1172–1178.

Erdely A, Hulderman T, Salmen R, Liston A, Zeidler-Erdely PC, Schwegler-Berry D, Castranova V, Koyama S, Kim Y-A, Endo M, Simeonova PP [2009]. Cross-talk between lung and systemic circulation during carbon nanotube respiratory exposure. Potential Biomarkers. Nano Letters 9(1):36–43.

Geiser M, Rothen-Rutishauser B, Kapp N, Schürch S, Kreyling W, Schulz H, Semmler M, Im Hof V, Heyder J, Gehr P [2005]. Ultrafine particles cross cellular membranes by nonphagocytic mechanisms in lungs and in cultured cells. Environ Health Perspect 113(11):1555–1560.

Göhler D, Stintz M, Hillemann L, Vorbau M [2010]. Characterization of nanoparticle release from surface coatings by the simulation of a sanding process. Ann Occup Hyg 54(6):615–624.

Harber P, Muranko H, Shvartsblat S, Solis S, Torossian A, Oren T [2003]. A triangulation approach to historical exposure assessment for the carbon black industry. J Occup Environ Med 45(2):131–143.

Harford AJ, Edwards JW, Priestly BG, Wright PFA [2007]. Current OHS best practices for the Australian nanotechnology industry. A position paper by the NanoSafe Australia Network. Melborne, Australia: The NanoSafe Australia Network.

He CN, Zhao NQ, Shi CS, Song SZ [2009]. Fabrication of carbon nanomaterials by chemical vapor deposition. Journal of Alloys and Compounds 484(1–2):6–11.

Holoman MN, Novotny M, Kemsley M, Raje J, Sullivan J, Pekarskaya T, et al. [2007]. The nanotech report. New York: Lux Research.

Huang RF, Wu YD, Chen HD, Chen C-C, Chen C-W, Chang C-P, Shih T-S [2007]. Development and evaluation of an air-curtain fume cabinet with considerations of its aerodynamics. Ann Occup Hyg 51(2):189–206.

ICRP [1994]. Human respiratory tract model for radiologic protection. ICRP Publication 66. Ottawa, Ontario, Canada: International Commission on Radiological Protection [http://www.elsevier.com/wps/find/bookdescription.cws_home/29164/description#description]. Date accessed: July 20, 2011.

Johnson AE, Fletcher B [1996]. The effect of operating conditions on fume cupboard containment. Safety Science 24(1):51–60.

Johnson DR, Methner MM, Kennedy AJ, Steevens JA [2010]. Potential for occupational exposure to engineered carbon-based nanomaterials in environmental laboratory studies. Environ Health Perspect 118(1):49–54.

Ju S-Y, Kopcha WP, Papadimitrakopoulos F [2009]. Brightly fluorescent single-walled carbon nanotubes via an oxygen-excluding surfactant organization. Science 323(5919):1319–1323.

Kane AB, Hurt RH [2008]. Nanotoxicology: the asbestos analogy revisited. Nat Nano 3(7):378–379, doi:10.1038/nnano.2008.182.

Karlsson HL, Gustafsson J, Cronholm P, Möller L [2009]. Size-dependent toxicity of metal oxide particles—a comparison between nano- and micrometer size. Toxicol Lett 188(2):112–118.

Kisin ER, Murray AR, Keane MJ, Shi XC, Schwegler-Berry D, Gorelik O, Arepalli S, Castranova V, Wallace WE, Kagan VE, Shvedova AA [2007]. Single-walled carbon nanotubes: Geno- and cytotoxic effects in lung fibroblast V79 cells. J Toxicol Environ Health Part A 70(24):2071–2079.

Lam CW, James JT, McCluskey R, Arepalli S, Hunter RL [2006]. A review of carbon nanotube toxicity and assessment of potential occupational and environmental health risks. Crit Rev Toxicol 36(3):189–217.

Lam CW, James JT, McCluskey R, Hunter RL [2004]. Pulmonary toxicity of single-wall carbon nanotubes in mice 7 and 90 days after intratracheal instillation. Toxicol Sci 77(1):126–34.

Li Z, Hulderman T, Salmen R, Chapman R, Leonard SS, Young SH, Shvedova A, Luster MI, Simeonova PP [2007]. Cardiovascular effects of pulmonary exposure to single-wall carbon nanotubes. Environ Health Perspect 115(3):377–382.

Lison D, Lardot C, Huaux F, Zanetti G, Fubini B [1997]. Influence of particle surface area on the toxicity of insoluble manganese dioxide dusts. Arch Toxicol 71(12):725–729.

Ma-Hock L, Treumann S, Strauss V, Brill S, Luizi F, Mertler M, Wiench K, Gamer AO, van Ravenzwaay B, Landsiedel R [2009]. Inhalation toxicity of multiwall carbon nanotubes in rats exposed for 3 months. Toxicol Sci 112(2):468–481.

Maynard AD, Kuempel ED [2005]. Airborne nanostructured particles and occupational health. J Nanopart Res 7(6):587–614.

Miyawaki J, Yudasaka M, Azami T, Kubo Y, Iijima S [2008]. Toxicity of single-walled carbon nanohorns. ACS Nano 2(2):213–226.

Murashov V, Howard J [2009]. Essential features for proactive risk management. Nat Nanotechnol 4(8):467–470.

Murray AK, Kisin E, Kommineni C, Kagen VE, Castranova V, Shvedova AA [2007]. Single-walled carbon nanotubes induce oxidative stress and inflammation in skin. Toxicologist 96(1):A1406.

Nemmar A, Hoet PHM, Thomeer M, Nemery B, Vanquickenborne B, Vanbilloen H, Mortelmans L, Hoylaerts MF, Verbruggen A, Dinsdale A [2002]. Passage of inhaled particles into the blood circulation in humans—response. Circulation 106(20):E141–E142.

NIOSH [1994]. NIOSH manual of analytical methods (NMAM®), Diesel particulate matter (supplement issues 3/15/03). 4th ed. By Schlecht PC, O'Conner PF, eds. Cincinnati, OH: U.S. Department of Health and Human Services, Centers for Disease Control and Prevention, National Institute for Occupational Safety and Health. DHHS (NIOSH) Publication No. 9–13 [www.cdc.gov/niosh/nmam/].

NIOSH [2005]. NIOSH respirator selection logic. Cincinnati, OH: U.S. Department of Health and Human Services, Centers for Disease Control and Prevention, National Institute for Occupational Safety and Health. DHHS (NIOSH) Publication No. 200–00 [http://www.cdc.gov/niosh/docs/2005-100/].

NIOSH [2006]. Criteria for a recommended standard: occupational exposure to refractory ceramic fibers. Cincinnati, OH: U.S. Department of Health and Human Services, Centers for Disease Control and Prevention, National Institute for Occupational Safety and Health. DHHS (NIOSH) Publication No. 200–23 [www.cdc.gov/niosh/docs/2006-123].

NIOSH [2007]. Progress toward safe nanotechnology in the workplace: a report from the NIOSH nanotechnology research center. Cincinnati, OH: U.S. Department of Health and Human Services, Centers for Disease Control and Prevention, National Institute for Occupational Safety and Health. DHHS (NIOSH) Publication No. 200–23 [www.cdc.gov/niosh/docs/2007-123].

NIOSH [2008]. Safe nanotechnology in the workplace. Cincinnati, OH: U.S. Department of Health and Human Services, Centers for Disease Control and Prevention, National Institute for Occupational Safety and Health. DHHS (NIOSH) Publication No. 200–12 [www.cdc.gov/niosh/docs/2008-112].

NIOSH [2009a]. Approaches to safe nanotechnology: managing the health and safety concerns associated with engineered nanomaterials. Cincinnati, OH: U.S. Department of Health and Human Services, Centers for Disease Control and Prevention, National Institute for Occupational Safety and Health. DHHS (NIOSH) Publication No. 200–25 [www.cdc.gov/niosh/docs/2009-125/pdfs/2009-125.pdf].

NIOSH [2009b]. Interim guidance for medical screening and hazard surveillance for workers potentially exposed to engineered nanoparticles. Cincinnati, OH: U.S. Department of Health and Human Services, Centers for Disease Control and Prevention, National Institute for Occupational Safety and Health. DHHS (NIOSH) Publication No. 200–16 [www.cdc.gov/niosh/docs/2009-116].

NIOSH [2009c]. Qualitative risk characterization and management of occupational hazards: control banding (CB). Cincinnati, OH: U.S. Department of Health and Human Services, Centers for Disease Control and Prevention, National Institute for Occupational Safety and Health. DHHS (NIOSH) Publication No. 200–52 [http://www.cdc.gov/niosh/docs/2009-152/].

NIOSH [2010a]. (Draft) NIOSH Current intelligence bulletin. Occupational exposure to carbon nanotubes and carbon nanofibers. Cincinnati, OH: U.S. Department of Health and Human Services, Centers for Disease Control and Prevention, National Institute for Occupational Safety and Health [http:www.cdc.gov/niosh/docket/review/docket161A/pdfs/carbonNanotubeCIB_PublicReviewOfDraft.pdf]. Date accessed: July 20, 2011.

NIOSH [2010b]. Prevention through Design. Cincinnati, OH: U.S. Department of Health and Human Services, Centers for Disease Control and Prevention, National Institute for Occupational Safety and Health [http://www.cdc.gov/niosh/topics/ptd/]. Date accessed: October 17, 2010.

NIOSH [2011]. Evaluation of health hazard and recommendations for occupational exposure to titanium dioxide. Cincinnati, OH: U.S. Department of Health and Human Services, Centers for Disease

Control and Prevention, National Institute for Occupational Safety and Health. NIOSH Publication No. 201–60 [www.cdc.gov/niosh/docs/2011-160/]. Date accessed: August 29, 2011.

NRC [2011]. Prudent practices in the laboratory: handling and management of chemical hazards (updated version). Washington, DC: National Research Council of the National Academies.

NSTC [2007]. The national nanotechnology initiative strategic plan. Washington, DC: National Science and Technology Council [http://www.nano.gov/NNI_Strategic_Plan_2007.pdf]. Date accessed: July 20, 2011.

Oberdörster G, Ferin J, Gelein R, Soderholm SC, Finkelstein J [1992]. Role of the alveolar macrophage in lung injury: studies with ultrafine particles. Environ Health Perspect *97*:193–199.

Oberdörster G, Ferin J, Lehnert BE [1994]. Correlation between particle size, in vivo particle persistence, and lung injury. Environ Health Perspect *102*(Suppl 5):173–179.

Oberdörster G, Maynard A, Donaldson K, Castranova V, Fitzpatrick J, Ausman K, Carter J, Karn B, Kreyling W, Lai D, Olin S, Monteiro-Riviere N, Warheit D, Yang H [2005]. Principles for characterizing the potential human health effects from exposure to nanomaterials: elements of a screening strategy. Part Fibre Toxicol *2*:8.

Oberdörster G, Sharp Z, Atudorei V, Elder A, Gelein R, Kreyling W, Cox C [2004]. Translocation of inhaled ultrafine particles to the brain. Inhal Toxicol *16*(6–7):437–445.

Oberdörster G, Sharp Z, Atudorei V, Elder A, Gelein R, Lunts A, Kreyling W, Cox C [2002]. Extrapulmonary translocation of ultrafine carbon particles following whole-body inhalation exposure of rats. J Toxicol Environ Health A *65*(20):1531–1543.

OECD [2002]. Frascati manual, 2002: proposed standard practice for surveys on research and experimental development, the measurement of scientific and technological activities. 6th ed. Paris: Organisation For Economic Co-operation and Development, p. 255.

OSHA [1990].29 CFR 1910.1450 Occupational exposure to hazardous chemicals in laboratories [www.osha.gov]. Washington, DC: Occupational Safety and Health Administration. Date accessed: July 20, 2011.

OSHA [1992]. General industry, safety and health standards. Washington, DC: Occupational Safety and Health Administration [http://www.osha.gov/pls/oshaweb/owadisp.show_document?p_id=12716&p_table=standards]. Date accessed: July 20, 2011.

OSHA [1994]. 29 CFR 1910.1200. Hazard communication, toxic and hazardous substances. Washington, DC: Occupational Safety and Health Administration. Date accessed: December 15, 2011.

OSHA [2008]. 29 CFR 1910.132. General requirements: personal protective equipment. Washington, DC: Occupational Safety and Health Administration. Date accessed: December 15, 2011.

Paik SY, Zalk DM, Swuste P [2008]. Application of a pilot control banding tool for risk level assessment and control of nanoparticle exposures. Ann Occup Hyg *52*(6):419–428.

Pauluhn J [2010]. Subchronic 13-week inhalation exposure of rats to multiwalled carbon nanotubes: toxic effects are determined by density of agglomerate structures, not fibrillar structures. Toxicol Sci *113*(1):226–242.

Poland CA, Duffin R, Kinloch I, Maynard A, Wallace WAH, Seaton A, Stone V, Brown S, MacNee W, Donaldson K [2008]. Carbon nanotubes introduced into the abdominal cavity of mice show asbestos-like pathogenicity in a pilot study. Nat Nanotechnol *3*(7):423–428.

Rao CN, Biswas K [2009]. Characterization of nanomaterials by physical methods. Annu Rev Anal Chem *2*:435–462.

Ryman-Rasmussen JP, Riviere JE, Monteiro-Riviere NA [2006]. Penetration of intact skin by quantum dots with diverse physicochemical properties. Toxicol Sci *91*(1):159–165.

Safe Work Australia [2012]. Safe handling and use of carbon nanotubes. Commonwealth Scientific and Industrial Research Organisation (CSIRO) and Safe Work Australia. ISBN 978-0-642-33351-3.

Sahu M, Biswas P [2010]. Size distributions of aerosols in an indoor environment with engineered nanoparticle synthesis reactors operating under different scenarios. J Nanopart Res *12*(3):1055–1064.

Samuneva B, Djambaski P, Kashchieva E, Chernev G, Salvado IMM, Fernandes MHV, Wu A [2006]. Sol gel synthesis and structure of hybrid nanomaterials with strong chemical bonds. In: Gdoutos EE, ed. Fracture of nano and engineering materials and structures. Dordrecht, The Netherlands: Springer, pp. 1037–1038.

Saunders GT [1993]. Laboratory fume hoods: a users manual. New York: John Wiley & Sons, Inc.

Schulte HF, Hyatt EC, Jordan HS, Mitchell RN [1954]. Evaluation of laboratory fume hoods. Am Ind Hyg Assoc Q 15(3):195–202.

Schulte P, Geraci C, Zumwalde R, Hoover M, Kuempel E [2008a]. Occupational risk management of engineered nanoparticles. J Occup Environ Hyg 5(4):239–249.

Schulte P, Rinehart R, Okun A, Geraci C, Heidel D [2008b]. National Prevention through Design Initiative. J Safety Research 39:115–121.

Seaton A, Tran L, Aitken R, Donaldson K [2010]. Nanoparticles, human health hazard and regulation. J R Soc Interface 7(Suppl 1):S119–129.

Shvedova AA, Castranova V, Kisin ER, Schwegler-Berry D, Murray AR, Gandelsman VZ, Maynard A, Baron P [2003]. Exposure to carbon nanotube material: assessment of nanotube cytotoxicity using human keratinocyte cells. J Toxicol Environ Health A 66(20):1909–1926.

Shvedova AA, Kisin E, Murray AR, Johnson VJ, Gorelik O, Arepalli S, Hubbs AF, Mercer RR, Keohavong P, Sussman N, Jin J, Yin J, Stone S, Chen BT, Deye G, Maynard A, Castranova V, Baron PA, Kagan VE [2008]. Inhalation vs. aspiration of single-walled carbon nanotubes in C57BL/6 mice: inflammation, fibrosis, oxidative stress, and mutagenesis. Am J Physiol Lung Cell Mol Physiol 295(4):L552–L565.

Shvedova AA, Kisin ER, Mercer R, Murray AR, Johnson VJ, Potapovich AI, Tyurina YY, Gorelik O, Arepalli S, Schwegler-Berry D, Hubbs AF, Antonini J, Evans DE, Ku B-K, Ramsey D, Maynard A, Kagan VE, Castranova V, Baron P [2005]. Unusual inflammatory and fibrogenic pulmonary responses to single-walled carbon nanotubes in mice. Am J Physiol Lung Cell Mol Physiol 289(5):L698–L708.

Shvedova AA, Kisin ER, Porter D, Schulte P, Kagan VE, Fadeel B, Castranova V [2009]. Mechanisms of pulmonary toxicity and medical applications of carbon nanotubes: two faces of Janus? Pharmacol Ther 121(2):192–204.

Singh N, Manshian B, Jenkins GJ, Griffiths SM, Williams PM, Maffeis TG, Wright CJ, Doak SH [2009]. NanoGenotoxicology: the DNA damaging potential of engineered nanomaterials. Biomaterials 30(23–24):3891–3914.

Takagi A, Hirose A, Nishimura T, Fukumori N, Ogata A, Ohashi N, Kitajima S, Kanno J [2008]. Induction of mesothelioma in p53+/- mouse by intraperitoneal application of multi-wall carbon nanotube. J Toxicol Sci 33(1):105–116.

Takenaka S, Karg E, Roth C, Schulz H, Ziesenis A, Heinzmann U, Schramel P, Heyder J [2001]. Pulmonary and systemic distribution of inhaled ultrafine silver particles in rats. Environ Health Perspect 1099(Suppl 4):547–551.

Tinkle SS, Antonini JM, Rich BA, Roberts JR, Salmen R, DePree K, Adkins EJ [2003]. Skin as a route of exposure and sensitization in chronic beryllium disease. Environ Health Perspect 111(9):1202–1208.

Tran C, Cullen R, Buchanan D, Jones A, Miller B, Searl A, Davis J, Donaldson K (1999). Investigation and prediction of pulmonary responses to dust, part II. In: Investigation into the pulmonary effects of low toxicity dusts. (No. 216/1999). Suffolk, UK: Health and Safety Executive.

Tran CL, Buchanan D, Cullen RT, Searl A, Jones AD, Donaldson K [2000]. Inhalation of poorly soluble particles. II. Influence of particle surface area on inflammation and clearance. Inhal Toxicol 12(12):1113–1126.

Trout DB, Schulte PA [2010]. Medical surveillance, exposure registries, and epidemiologic research for workers exposed to nanomaterials. Toxicology 269(2–3):128–135.

Tsai S-J, Ada E, Isaacs J, Ellenbecker M [2009a]. Airborne nanoparticle exposures associated with the manual handling of nanoalumina and nanosilver in fume hoods. J Nanopart Res 11(1):147–161.

Tsai S-J, Hofmann M, Hallock M, Ada E, Kong J, Ellenbecker M [2009b]. Characterization and evaluation of nanoparticle release during the synthesis of single-walled and multiwalled carbon nanotubes by chemical vapor deposition. Environ Sci Technol *43*(15):6017–6023.

Tsai S-J, Huang RF, Ellenbecker MJ [2010]. Airborne nanoparticle exposures while using constant-flow, constant-velocity, and air-curtain-isolated fume hoods. Ann Occup Hyg *54*(1):78–87.

UNH [2009]. Nanomaterials safety program. Durham, NH: University of New Hampshire [http://www.unh.edu/research/sites/unh.edu.research/files/docs/EHS/Chem-safety/UNH-Nanomaterials-Safety-Program.pdf]. Date accessed: July 20, 2011.

UNC [2011]. Summary of recommended nanomaterial risk levels (NRL). Chapel Hill, NC: University of North Carolina at Chapel Hill [http://ehs.unc.edu/ih/lab/nano.shtml]. Date accessed: July 20, 2011.

Wagner GF, Fine LJ [2008]. Surveillance and health screening in occupational health. In: Wallace R, ed., Maxcy-Rosenau-Last. Public health and preventive medicine. 15th ed. New York: McGraw-Hill Medical Publishing, pp. 759–793.

Zalk D, Paik S, Swuste P [2009]. Evaluating the control banding nanotool: a qualitative risk assessment method for controlling nanoparticle exposures. J Nanopart Res *11*(7):1685–1704, doi: 10.1007/s11051-009-9678-y.

www.ingramcontent.com/pod-product-compliance
Lightning Source LLC
Chambersburg PA
CBHW081901170526
45167CB00007B/3106